Stephanie Wakefield

Anthropocene Back Loop

Experimentation in Unsafe Operating Space

CCC2 Irreversibility

Series Editors: Tom Cohen and Claire Colebrook

The second phase of 'the Anthropocene,' takes hold as tipping points speculated over in 'Anthropocene 1.0' click into place to retire the speculative bubble of "Anthropocene Talk". Temporalities are dispersed, the memes of 'globalization' revoked. A broad drift into a de facto era of managed extinction events dawns. With this acceleration from the speculative into the material orders, a factor without a means of expression emerges: climate panic.

Stephanie Wakefield

Anthropocene Back Loop
Experimentation in Unsafe Operating Space

()
OPEN HUMANITIES PRESS
London 2020

First edition published by Open Humanities Press 2020

Copyright © 2020 Stephanie Wakefield

Freely available at:
http://openhumanitiespress.org/books/titles/anthropocene-backloop

This is an open access book, licensed under Creative Commons By Attribution Share Alike license. Under this license, authors allow anyone to download, reuse, reprint, modify, distribute, and/or copy their work so long as the authors and source are cited and resulting derivative works are licensed under the same or similar license. No permission is required from the authors or the publisher. Statutory fair use and other rights are in no way affected by the above. Read more about the license at creativecommons.org/licenses/by-sa/4.0

Figures, and other media included with this book may be under different copyright restrictions.

Cover illustration by Caroline Castro. Copyright 2020, CC-BY-SA 4.0

Print ISBN 978-1-78542-071-9
PDF ISBN 978-1-78542-072-6

OPEN HUMANITIES PRESS

Open Humanities Press is an international, scholar-led open access publishing collective whose mission is to make leading works of contemporary critical thought freely available worldwide. More at http://openhumanitiespress.org

Contents

Prelude   7

Acknowledgements   8

Introduction   10
   Why back loop?
   Responses to the back loop and experimentation (overview)
   Nobody knows what this is
   Free to move on other planes

1   **The Back Loop**   20
   The coordinates are scrambled
   Adaptive cycle
   Stitching the whole Earth together (Anthropocene front loop)
   Splintering whole Earth (Anthropocene back loop)
   Now what?

2   **Government in the Back Loop:**
    Resilience and Managing Safe Operating Space   34
   Governing safe operating space
   Governing urban risk, front to back loop
   New York City as lab
   Redefining infrastructure
   Further south in Miami
   Resilience's relationship to the back loop
   Experimentation and power

3   **Sit Down, Be Humble: Imaginaries of**
    Post-Apocalyptic Survival Amidst Interlinked Ruins   56
   Imagining life beyond safe operating space
   Sovereign grounds and hatred of the self
   Earthbound
   A new sentimental education
   Ruins theory
   It's beautiful beautiful beautiful
   The Anthropocene moral code
   Dwelling in back loop ruins
   The power of mental imagery at the end of the world
   Love is also a ground

	Interlude: Getting Out of the Loop	80
4	**Survival Skills and Floating Houses**	85
	Disaster prepping and survival skills	
	Okay, why am I working for that guy?	
	Living with water	
	Infrastructure	
5	**Use of the Body**	103
	Let the bodies hit the floor	
	Looting of practical and theoretical stores across fitness and sport	
	Independent living	
	Reinventing mind and body	
6	**Use of Our Soul**	114
	Roots revival?	
	Politics is like for 2,000 years ago	
	Wherever our most high leads	
	Mi dweet fi di love mi nuh dweet fi di likes	
7	**Out of the Back Loop**	124
	Experimentation in unsafe operating space	
	Free use	
	Tools of freedom	
	Autonomy	
	Plus ultra	
	But the Earth does not dream of you	
	Into the unknown	
	Coda	142
	Notes	144
	Works Cited	181

Prelude

This is the doomsday future we are offered. As oceans rise and temperatures soar, coastal cities are inundated, and swathes of the Earth become uninhabitable. In Miami where I currently live, one foot of sea level rise—expected by 2040—leads saltwater to infiltrate Biscayne Aquifer, rendering tap water undrinkable and devastating the sewage system. Residents flee en masse like 21st-century Dust Bowl refugees. For those who stay, instead of bikini shops and ventanitas, an increasingly-submerged, unbearably hot city of undrivable highways surrounded by virus and excrement-filled oceans.

In most depictions of our future, life for ordinary people is diminished, reduced to survival and precarious getting by. Such imaginaries are common. But despite their ubiquity, are these images of life the only ones possible?

Acknowledgements

Writing is a dialogue between the author and the multifarious realms of life experience, dreams, and personal interaction. On the latter, there are many people I want to thank for their support, advice, and conversation. I am extremely grateful to the editors of the Critical Climate Chaos: Irreversibility series, Claire Colebrook and Tom Cohen, for their support and feedback on this book. Thank you to Bruce Braun, for so much support and friendship over many years. Hutch Ismael Valentin, Janice Vargas, Estefania Panesso, Julian Chupacabra Monar, Frankie Cruz Echevarria, Peppa Holder, Mia Matković, Dominika Jargilo, Prabesh Gurung, and everyone else from CrossFit Breed: your friendship changed my life and I would not be the same without you. Kevin Grove, for extensive support on the way to and at Florida International University. Elizabeth Johnson, who introduced me to CrossFit. Caroline Busta, whose brain is always five steps ahead and continuously motivates me to push my own further. David Chandler, for conversations and especially critiques. Nigel Clark, for lots of inspiration and the idea of "unsafe operating space." Clark Fitzgerald, Kat Nelson, Corey Eastwood, Cosmo Bjorkenheim, Elise Duryee-Browner, Alyssa Bonilla, Tony Buczko, Antek Walczak, Gean Moreno, George Sanchez Calderon, Elizabeth English, Jacques Lacour, and Buddy Blalock.

The book was written primarily while I was visiting Eugene Lang College-The New School as an assistant professor of Culture and Media. There, I would especially like to thank McKenzie Wark for her incredible support, as well as Shannon Mattern, Dominic Pettman, Simon Critchley, Timon McPhearson, and Ed Keller. I am grateful also to my students at Queens College and The New School, especially those in my Anthropocene Futures, Anthropocene Life, and Infrastructure courses. Particular thanks to Annabel Finkel, Galen Peterson, Stockton Cobb, Jacob Channel, Isabella Kazanecki, Jil Averbeck, Kate Bilezikian, Jessica Davies, and Dawit Yohannes Kiflemariam. You inspired parts of this book and I hope you will find yourselves present in it.

The final writing and editing of the book were made possible by a generous fellowship from the Urban Studies Foundation. I am very grateful for this support.

Parts of the book were presented at meetings of the American Association of Geographers in Boston and New Orleans; the Design

4 This Century seminars at The New School; the Miami Beach Urban Studies seminar at Florida International University; the "Anticipation, Experimentation & Design in the Anthropocene: On Governing Futures of Life" Research in Conversation Workshop at Durham University; the Urban Studies Foundation Symposium at the University of Glasgow, the Department of Global and Sociocultural Studies SAGGSA Colloquium Series at Florida International University, Post-Planetary Futures symposium at The New School, and the Measuring the Anthropocene Symposium at the Center for Humanities and Information at Pennsylvania State University. For invitations to speak at these events and/or for thought-provoking discussion at them, huge thanks to Oliver Belcher, Claire Colebrook, Melanie Crean, Kevin Grove, Marisa Jahn, Elizabeth Johnson, Ed Keller, Jonathan Pugh, and Kathryn Yusoff.

Parts of the book or early versions of ideas developed in it were previously published in different forms in "Amphibious Architecture Beyond the Levees," *Mobilities* (2019); "Infrastructures of Liberal Life: From Modernity and Progress to Resilience and Ruins," *Geography Compass*, 12, no. 7 (2018); "Inhabiting the Anthropocene Back Loop," *Resilience: International Policies, Practices and Discourses*, 6, no. 1 (2018): 1-18; Stephanie Wakefield and Gean Moreno, "Tools for a Thawing World." Exhibition catalogue essay for SUPERFLEX: We Are All in the Same Boat, Museum of Art and Design, Miami Dade College (2018). "Dreaming the Back Loop," *Affidavit* (2017); "Field Notes from the Anthropocene: Living in the Back Loop," *The Brooklyn Rail* (2017). Thank you to Hunter Braithwaite and Paul Mattick for generous feedback on these last two pieces.

A special thank you to KD, who kept me company while I finished editing the book. Most importantly, thank you to GD. I love you so much.

Stephanie Wakefield
March, 2019

Introduction

Why back loop?

Liberalism's old structures are unraveling. We are free to create our own. That is the basic premise of this book.

When the Anthropocene first entered public discourse around 2011, it seemed at least to my mind to offer an opening, a way to break out of ineffectual political frameworks, and to take up the basic matters of transforming what it means to be alive, within social and political flux. For this reason, I experimented with it in writing and in political practice as a name for the present. Over the years however the term been molded and modulated in many ways by resilience proponents as much as by critical theorists, across whose diverse discourses it has become nearly synonymous with entanglement, antihumanism, and diminished expectations. Over and over, when the word Anthropocene is uttered, it is followed by listicles of environmental destruction or musings on the scale and scope of crisis, which, rather than inciting any range of actual responses, themselves lead to moralizing instructions on the necessity of educating the masses, using smarter lightbulbs, jettisoning "outdated" nature/human binaries, and supporting resilience gurus or a green new politician. In so far as it is perceived as such, the Anthropocene has become a hindrance, rather than an opening, to transforming life. I am exhausted by the apocalyptic and hateful images being forced on us by political culture in this and all regards, and by the rigid modes of discourse that now portray life in the Anthropocene as survival amid entangled ruins of a broken world. "Annihilation! Annihilation! Annihilation!" the psychologist in recent cli-fi book *Area X* screams repeatedly, hoping to make the biologist self-destruct, only to realize that the more she repeated it, the more meaningless the word seemed. Like apocalypse, the End, or all the *Guardian Environment* articles retweeted each day—the word tends to lose its power with aggravated repetition.

This book proposes an alternative reading of the present, one which seeks to break with these crisis-ridden contemporary imaginaries, and to again see other possibilities open now. To do so I borrow the heuristic of the back loop from resilience ecology. As explained in chapter one, a back loop is a time of confusion and collapse as well as potential and reorganization. I find the back loop useful as a lens through which to see the present because it allows phenomena to appear not as endless crises or self-confirming signs of impending doom, which ultimately only shore up one's own preexisting beliefs, but instead as the singular responses they are in a liberal society in freefall.

I started thinking about the back loop in the spring of 2017 while teaching resilience ecology at The New School in New York. Despite the fact that my research on resilience has always been extremely critical, the concept of the back loop struck me as a compelling way to think about the moment we're living in at society or civilizational level. At the time I was experiencing many deep transformations in my own personal life as well, with a lot of the basic assumptions and modes I'd been living in being upended. In so far as it implies responding to such situations at whatever scale by allowing oneself to let go and actually experience them, allowing metamorphosis to occur rather than holding on to old frameworks senselessly, the back loop concept made sense to me on that level as well. In general, what I seek to do with the back loop, as I'd tried to previously with the Anthropocene, is to get away from the really unimaginative frameworks used in politics to understand life, and to test a way of seeing that would break with preestablished categories. Such approaches assume that life is a known quantity, reducing it either to surviving impending apocalypse or a range of preexisting yet completely inadequate political models. Instead what happens if we look at the changes happening around us from a less rigid or calcified perspective, and instead see that while one set of codes for living are coming undone, this doesn't need be a tragedy? It can also mean that we have an opportunity, one many people have actually wanted for a long time, to create our own new codes, now.

To be clear, by borrowing the back loop concept from resilience ecology, I am not importing the field's attendant frameworks wholesale or accepting resilience's view of the world in terms of systems. Such would accept a cybernetic view of life, which I neither find sufficient nor accurate. Instead I am borrowing the back loop heuristic as one tool among countless others for trying to think the now. As I will suggest toward the book's end, inhabiting the back loop may lead to the end of any such loops altogether. Likewise, the goal in using this heuristic is not to own or constrain the present. It is not to brand it, nor to try to define it once and for all. Instead it is just one provocation to thought, a proposal to shift out of established liberal models that currently dominate. A lens through which to see contemporary phenomena differently. There could and should be many other ways to think the present. The goal of this book is simply to put the idea of the back loop and contrasting ways of responding to it on the table for discussion and elaboration.

Responses to the back loop and experimentation (overview)

Times of deep change are usually perceived existentially rather than cerebrally—singularly rather than abstractly. As such this book explores the back loop beginning from what is being done—from practices occurring now. Practices of power and truth, dreaming and living, governing and shaping: such practices are as old as humans themselves. They are how we create our worlds, take them in hand and shape them. But what is happening to these practices of life as they enter the back loop?

The book is a tour through some of the back loop's linked iterations, exploring the different ways different people are responding to and living in it, as well as some lessons that we can learn from their diverse practices. Obviously, the ones I explore are only some amidst a broader range of practices. Not everyone experiences the back loop in the same manner (the desire to go back is equally, if not more, ubiquitous than the desire to inhabit here). That said this book is for the new. My interest is in practices that directly take up back loop dislocations in their own unique ways. This holds for "malevolent" and "liberatory" practices alike, and includes the activities of governments, technologies, and ordinary people within back loop transformations. But throughout my concern is to explore possibilities of liberation and freedom for ordinary people in a back loop context, and the way in which these possibilities can be redefined within its shifting configurations.

One way to respond to the back loop is to try to maintain safe operating space via deploying new modes of management. I discuss two versions of this response in chapters two and three. Chapter two explores one of the most dominant ways of responding to the back loop: resilience. Resilience is the current incarnation of liberal governance, a transfiguration in its own modes and techniques. Defined as "the ability of a system to absorb disturbance and still retain its basic function and structure,"[1] resilience has risen to the top of urban management agendas, replacing sustainability as means, end, and theoretical framework. Moving between reflections on my research on resilience infrastructures in New York and Miami and theoretical and cultural site-based analysis, this chapter analyzes the political, social, and technical dimensions of urban resilience and its hopes of preserving and managing global urban systems in their "safe operating space." I argue that although many celebrate resilience as the city or planet's salvation, it supports a disabling fiction whereby human survival in the era of climate change is tethered to the maintenance of existing economic, social, and political relations. In contrast to management past, which promised a better future, the role of resilience technologies in coastal cities, I argue, is to manage and adapt to changing conditions of catastrophe at sea—rising seas and storm surge—in order to secure and manage an unchanging urban order on land.

In chapter three I trace the affinity between resilience's desire to maintain the old operating space and imaginaries of post-apocalyptic life present in contemporary critical theory and media, with special focus on recent climate fiction and theoretical work of Bruno Latour, one of the most widely heard voices of the Anthropocene. Surveying speculative imaginaries of post-apocalyptic life found in film, fiction, and critical theory, I argue that what is emerging are not novel imaginaries that help us generate other possible ways of living—as I believe many in this field hope. Instead their works impose a new "Anthropocene moral code" with damaging constraints on life and imagination, demanding adherence to laws of entanglement, antihumanism, and limits, with failure to do so risking being seen as obsolete hubris. As such, I argue, these imaginaries aim to stabilize (to govern) life, albeit by declaring the latter to be unstable and outside human control.

I have another major disagreement with this body of literature. To be very clear, the "front loop" I describe in this book is not "our" world for which we are all equally responsible; it is the world forwarded and enforced by liberal regimes. I am not in agreement with other Anthropocene theorists for whom "our" world is ending, one which "we" profited from. From my vantage point the world of liberalism must be understood as an historically specific regime of government, with interests in discipline, profit, and productivity, which sought to mold much of life, human and nonhuman, in its image. Another way to describe the back loop is to say that society is witnessing the slow motion trainwreck of this regime in its last moments, and we are its collateral damage. In this context both resilience and ruins imaginaries impose a blackmail of a single loop on all of us. Human survival and the survival of the liberal way of life are conflated into one. We are to choose between a life of governance or a life of governance.

In chapters four, five, and six I explore other ways people are responding to the back loop. In contrast to resilience's nihilistic reduction of life to crisis management, and in further contrast to critical accounts of the Anthropocene that celebrate the life of things or a world without humans, these short chapters show that our time is just as equally one of great experimentation in human capacities and ways of living. Chapter four deals with skills and infrastructure, exploring ways in which people across the US are taking up survival skills, technologies, or amphibious architecture and, through these, the means of existence within shifting environments. Chapter five moves to the scale of the body, to look at how contemporary physical fitness movements, in particular CrossFit, are developing new physical practices by looting existing stores of fitness regimes, and in process taking possession of health and body amidst a tidal wave of chronic disease and decrepitude. Chapter six is concerned with possibilities for crafting new forms of subjectivity in the back loop, and looks at the work of contemporary Jamaican singer Chronixx and possibilities present, as he puts it, in experimentation with one's soul. These three chapters are short and focused on extremely specific examples, some of which may resonate more strongly with readers than others depending on their sensibilities. What these chapters show are other responses to the back loop by regular people, which they develop not to govern the back loop, but to inhabit it according to their own needs and inclinations. Moreover, they seek to show the vast range of valences within which it is possible now to rethink life and how it is lived. Together these chapters ultimately argue that living in the back loop requires a new practical orientation, a letting go of old frameworks, hubristic experimentation with new uses, and an allowance for the unknown—and all of this imbued with a confidence in one's own pathways. In the use of tools, use of the body, and use of one's own soul, we are witness to a deeply dramatic taking up of life—how it is described, defined, and lived.

Though such practices are extremely diverse, they can also illuminate a divergent way of responding to the back loop, which, rather than governing it, concerns living with autonomy in and beyond it. To explore this in the final chapter I reevaluate and employ theoretical tools found in the works of Michel Foucault, Giorgio Agamben, and Peter Sloterdijk, as well as methodologies derived from ecology, to theorize these practices, as a means of understanding how people can appropriate and transform their worlds. In contrast to resilience's efforts to preserve the old front loop safe operating space, this chapter argues for a widespread, popular taking up of the possibilities of the back loop by experimenting with one's own modes of inhabiting it. This I suggest requires transfiguration and reclamation of tools of existence and the hubristic confidence to wield them. In keeping with the speculative and experimental spirit I have outlined, my goal is to put such ideas on the table, so that they may be discussed, debated, explored or rejected. Embracing experimentation in the back loop could lead us to unpredictable, provisional collaborations and divergences whose outcome cannot be known or predicted in advance. Much remains to be explored.

As a whole the book works as a tour through landscapes of the back loop opened by each of these different responses. Each chapter can be seen as a snapshot or vignette of a different trajectory in it. As it proceeds the book shifts registers and styles as much as scales, to track and develop on some of these pulsations. This is intentional and designed to evoke different frequencies on which the back loop is experienced. These iterations can be read as linked or coexistent, potentially feeding or limiting one another, but they may also be seen as divergent trajectories that delink and take on their own independent velocities. Each represents both broader paradigms of back loop response, while also maintaining their own singular rhythm as practices.

In the book I use experimentation as the best word I could find to describe both practices of back loop governance and for living with more freedom and autonomy in the back loop. I do so not to suggest a grand new concept of experimentation,[2] but just to note that, amidst back loop dislocations and shifts, diverse people and forces are jettisoning preexisting models and experimenting with new ones. In the realm of governance, one finds real-time experiments in urban environments conceived as living laboratories, with the aim of managing urban populations and environments and, as I argue, more fundamentally maintaining liberal orders by new means, projecting them into the future infinitely even amidst the catastrophes they generate. In the case of resilience, these experimental practices often result from a perceived obligation to relinquish past modes of urban administration and to embrace a new ecocybernetic approach. On the other hand, practices I discuss in the second half of the book in terms of autonomy may be described as experimental in the sense that they do not follow from exterior political or moral blueprints; instead emerge from within the needs, lives, and dreams of practitioners themselves; are enacted and made use of by practitioners themselves; are modulated over time as practitioners discover new needs, desires, or limits to overcome; do not seek as their end a specific society or scenario, but are better described as tools for living, for taking one's life into one's own hands and in so doing making it into a work of art; and which finally open onto possibilities unpredictable in advance, and see this as a fine thing. In this sense there is a freedom to them. While there are perhaps many formal similarities between the experimentation of resilience government and that of autonomously inhabiting the back loop—one may also easily remark that experimentation broadly speaking is a liberal category—there are also important differences. While resilience seeks to *govern* back loop dislocations, practices described in the book's second half seek variously to freely inhabit the back loop, and to take their lives in hand within it in various shades and tenors. Focusing exclusively on formal qualities of experimentation may lead to readers missing this more fundamental difference of sensibility.

Nobody knows what this is

While for the purposes of the book I limited myself to these examples, I could have easily looked at many other trajectories and examples. I could have, for example, also discussed the unpredictable rise of post-truth itself as a back loop phenomenon, that is, the idea of a single objective truth, to which you could speak truth to power and it would all be revealed, shattered by the recent agreement that most media images are and have always been fabrications, and by a man who just realized you can say whatever you want, not apologize for it, and suddenly everyone feels entitled to their own definition of truth. Other frequencies of the back loop that could be discussed might include the seeking online and IRL of new ways to understand the world now manifesting variously on social media, or the rise of movements from Occupy to Gilet Jaunes for which "nobody knows what this is" has been as much an assessment as it has been a rallying cry and badge of honor. On the rise of experimentation as a back loop mode of governance and flight both, a second volume of this book might cover developments in the realm of space colonization such as SpaceX, blockchain, or Google's desire to aggregate human being into data. Together and in their conflict these practices make up the back loop not as a thing or universal encompassing epoch, but an epochal assemblage with many possible trajectories.

Just as there is not one truth of the back loop, equally there is no single truth to any of these stories. There are always many angles, and many ways in which what seems negative from one angle can become useful or take other directions from another. Thus chapters two and three, which are primarily critical, conclude with thresholds, points where the topic under discussion exceeds itself, spiraling beyond even its own goals toward other horizons (the element most worth finding in any situation). I finish chapter two, for example, with the argument that resilient urbanism, despite its problems, offers tools for a different way of responding our turbulent epoch, in the form of the experimental ethos that suffuses its efforts. Just as equally, if the back loop premise proves useful to readers, it will hopefully do so by leading them in completely different directions.

Truth is ultimately a local and contingent thing. Not only does this perspective entail recognition that that which we are told represents absolute fact—as portrayed by CNN or Fox, for example—is in reality the interest of this or that party. Moreover this perspective on truth entails understanding truth as an extremely personal matter. We have always been post-truth. This does not mean that there is no truth, but that there are as many truths as there are peoples, dreams, and realities. Following this perspective, this book does not aim to get the truth of the back loop. It aims rather to insist that there *isn't* one. In arguing such, it is engaged in what Foucault once called a "battle 'for truth', or at least 'around truth',"which, he explained, is not for an absolute truth can be "discovered and accepted," but is a battle about "the rules according to which the true and false are separated and specific effects of power attached to the true."[3] In contrast to efforts to lay down the meaning of the Anthropocene for everyone, the book instead argues that there are many truths to the back loop, many ways in which it is experienced, understood, and taken up. Experimental techniques of resilience government, techniques of imagining post-apocalyptic life, and techniques of free inhabitation: each are frequencies of the back loop, different ways of responding to the questions and problems it poses, is seen to pose, uniquely across place and time. Each create and follow their own forms of knowledge, practices of truth, and technologies of power and life, as well as ways of constituting subjectivity on the basis of these imaginaries and practices.

What all of these trips through the back loop and its iterations ultimately suggest is that we are heading not into a single imagined future but that the loop, long imagined in the singular, is spiraling out beyond its own bounds. Decoupling, breakaway subjects, or even breakaway civilizations—it remains to be worked out.

Free to move on other planes

In the end, I do not offer a solution to the back loop, at least not in the traditional definition of a problem-solving response, a means to a final resting place without conflict or change. In my view the back loop is not a problem to be solved. Rather, the shift I am outlining in this book is simply toward a different thinking about transformation and about life. To say there is no solution is not the same as saying, as do governments and theorists, that nothing else is possible, so we may as well keep greasing the wheels along this catastrophic path. To say there is no solution is rather to suggest that this calcified way of thinking is part of the problem. A lesson I take away from the experiences I have had over the past decade in political movements such as Occupy, as well as in experiments which emerged from such moments based on the Anthropocene hypothesis—during which the question of why political modes of thinking and acting are so limited, why despite their stated aims of transforming the world anew they instead contribute to its maintenance and the dulling of imagination within it reappeared to me incessantly—is not that past forms were wrong and what is needed is to find the next new correct one. The lesson is that the one form that would finally end all conflict, finally establish paradise on Earth, will never be found. Moreover, that spending our whole lives searching in vain to establish it leads us to miss what is already possible here, now. My suggestion that we are in the back loop means that we have already crossed various tipping points, but that in doing so, everything from social practices, technologies, and truth to plants, animals, and places have become shaken out of their normal frameworks. We are free to move on new planes. And this should compel us to shift our perspective a bit. Instead of lugging around old political frameworks for no reason, and then trying in vain to make our new realities conform to them, we now have the opportunity to stake out entirely new possibilities for ourselves and each other. I do not think we require alternative ways of organizing the whole world or society, but instead to take back the conditions for asking what life can be. We likewise do not need new laws, but instead new practices and tools for getting on with the immeasurable beauty that is living, for giving new sense to the secure, happy, rich, or meaningful life. Instead of looking for solutions *to* the back loop, it seems more relevant to explore the possibilities offered to us *by* it, the potential it holds here and now. What's on the table now is what is true, what being human means, as well as who gets to answer those questions and in what ways.

In the back loop, we have the opportunity to rethink, define, and powerfully shape our existence. From this perspective our time is a time for audacity, experiments on the same playing field where our future is already being written for us.

In terms of scale, many practices are going to be hyperlocal—they might not make sense in other environments—while others are undoubtedly going to require complex global cooperation. A whole civilization in the back loop. This is a venture open to all of us. Fundamentally this is going to be a vast experiment, wildly imaginative and deeply democratic in nature. It's something that has to happen but something that many of us also really want. After all, one of the only things most Americans agree on is that, at a bare minimum, we need a revolution.

The changes we need are going to be deeply transformative—going for them will require both daring and courage—but that doesn't mean they won't draw on what exists now or traditional practices and behaviors. It's not about the new or old, but about new combinations and arrangements for a meaningful and livable existence on this planet. This orientation entails finding new modes of nourishing ourselves, designing and raising buildings, staying warm or cool, and accessing clean water as it is does learning to face the unknown and learning to look into ourselves and ask what kind of life *we* want to make live, what kind of life is worth living, and asking previously unaskable questions. What is being human? By "we" I don't just mean designers, city governments, planners, or resilience theorists who have already become back loop participants, as testified by the existence and growth of the resilience paradigm. Nor do I mean a fictional homogeneous "we" of the species, the assumed liberal body social, or any other. By "we" I mean anyone: common people where they are, how they are, people who will bear the brunt of climate change, people who already needed the world to end yesterday.

1 The Back Loop

The coordinates are scrambled

Record heat waves blaze across Europe and North Africa; wildfires scorch Greece and even the Arctic Circle, while on America's West Coast, meteorologists warn that billowing smoke could choke out views of annual Perseid meteors. According to Earth systems scientists, the planet is shifting out of the stable climates of the 11,000-year-long Holocene interglacial in which modern civilizations developed (and also out of the glacial-interglacial cycle in which it flickered for the last 100,000 years) and into the Anthropocene, a more volatile and unknown operating space where the glaciers are melting, seas are rising, and climates are changing.[4] Fueled by carbon dioxide emissions and biosphere degradation, this headlong movement will surpass various tipping points of biogeophysical feedback—permafrost disappearance, land and ocean carbon sinks weakening, polar ice sheet melting—accelerating global warming as well as pathway irreversibility.[5] While for some scientists the term is a matter of the impact humans will leave on Earth in the deep geological future and is thus a matter of geology, for other Earth systems scientists such as Owen Gaffney and Will Steffen, the Anthropocene represents "humanity's effect on the Earth cross[ing] a tipping point" and the Earth's shifting into a new domain of operation, out of the Holocene's safe operating space and into a question mark, with Earth heading into "planetary terra incognita."[6] In this sense the Anthropocene offers a name for a time of profound transformation.

The boundary crossings and dislocations of the present also concern the human realms of thought and action. Alongside the Earth's transformations, we are equally contemporaries of liberal civilization's shifting baseline.[7] While interpretations are diverse, it is clear is that in the Anthropocene the grounds, parameters and imaginaries for thought and life are being upended and shook loose from their moorings. We witness this dislocation in what is seen by many as the bypassing, scrambling or breaking down of modernity's unified categories of Human or Nature said to mark the Anthropocene, transfigurations leading new visions of human life to emerge. We see it equally in the unexpected rise of the post-truth age and the "challenging [of] well-established dualistic boundaries such as nature and culture or good and bad," as scientists Jan Zalasiewicz, Will Steffen, Reinhold Leinfelder, Mark Williams and Colin Waters put it.[8] Likewise

The Back Loop

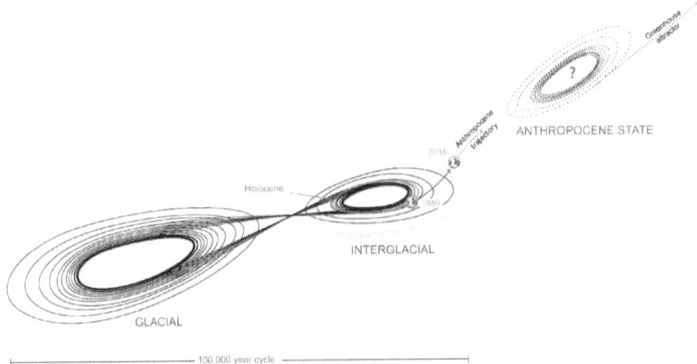

Figure 1.1: The trajectory of the Anthropocene, showing Earth beginning to move out of its glacial-interglacial cycles of the late Quaternary in 2016 and onto a new, unknown pathway. Adapted from Owen Gaffney and Will Steffen, "The Anthropocene Equation," *The Anthropocene Review*, 4, 1 (2017): 53–61. Reprinted by Permission of SAGE Publications, Ltd.

geopolitically we see Anthropocene destabilization in regimes spending millions of dollars trying to ward off crisis; the United States and the United Kingdom's attempting exiting of liberal global order; the giving way of old modes of social regulation to the soft bans and behavior policing of tech companies like Facebook and Google; while at other scales and in various sites, new tribalisms of all stripes develop.[9] Other times we sense present dislocations in less cerebral manners. In so many ways, as artist Gean Moreno has put it, "we can feel—the modulations are like soft electrodermal pulsations, but on the inside—our imaginary undergoing significant restructuring."[10]

Adaptive cycle

How to think this situation? For some critical thinkers, the present is experienced as something to lament, a catastrophe to endure, or crisis to manage. For others we are living in an apocalyptic end time, either literally or in its temporality.[11] Many now celebrate the end of Man and the "life of things" or the "world without us," offering new antihumanisms appropriate to a moment in which our potential extinction is much discussed.[12] But to better capture the complex depths and textures of our time—without nostalgia or morality—rather than a crisis or the end of the world, I argue that we are living in not only the Anthropocene but also, more specifically, its "back loop," a time of release, fragmentation, and great potential for reorientation.[13] The back loop concept was developed by C.S. Holling in the 1970s and

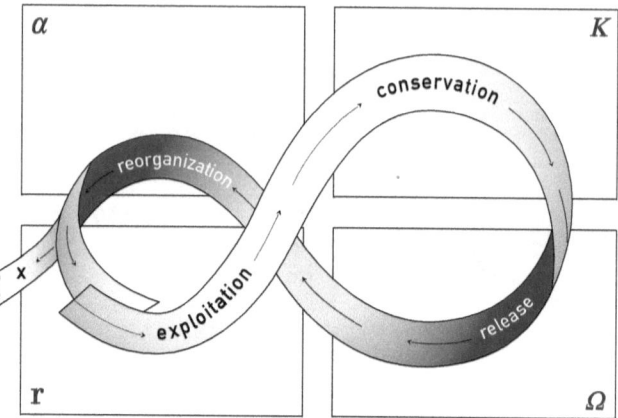

Figure 1.2: Adaptive cycle. Design adapted from Lance H. Gunderson and C.S. Holling, Panarchy: Understanding transformations in systems of humans and nature (Washington, D.C.: Island Press, 2002), p. 34, by Caroline Castro for this book, 2019.

is now used widely by resilience ecologists.[14] For ecologists, every system (a forest, a body, a city, etc.) goes through a cycle with two phases, what they call a "front loop" and a "back loop" (together making up the "adaptive cycle" (see Figure 1.2)).[15] The front loop is seen as progressing from an initial growth or exploitation phase to conservation and seeming stability, while in a back loop those structures come apart, leading to a period of destabilization, fragmentation, confusion, and release but also great potential for experimentation, reorganization, and transformation.[16]

The back loop is a relatively new and little-studied concept. Until the 1970s, ecologists viewed ecosystems through a teleological model of succession, seen as progressing from the initial growth or exploitation phase, represented by r, to a second and final phase of stability represented as K.[17] To take the classic forest example, the first phase is dominated by fast-reproducing pioneer species that colonize and exploit a fresh base of abundant resources.[18] Over time, they are replaced by larger, more specialized organisms, which annex the system's niches and nutrients. The result is a mature forest, a stable, tightly connected climax community where everything—sunlight, water, biomass—is "in its place."[19] The climax phase was viewed thus as the ideal end point, where a system's steady state was made up of the organisms best adapted to its environment. Equilibrium was the key idea for systems understood to possess homeostatic modes (homeotherms, etc.). For most of ecology's history, environmental management was geared towards conserving and managing ecosystems in this stage. They thought, in other words, the front loop was all there was to life.

This model underwent revision in the 1970s when Canadian ecologist C. S. Holling made his now-well-known interventions that led to the new field of resilience theory.[20] Systems, Holling argued, do not remain in a single steady state. Rather, they regularly experience phases of release and reorganization, times of collapse, creative destruction and renewal. By comparing myriad case studies of diverse ecosystems, Holling and colleagues argued that it was necessary to add another loop, a so-called back loop. For resilience ecologists, back loops usually occur due to a sudden crisis event: forest fire, flood or pest outbreak.[21] In the release phase—represented by Ω— energies and elements previously captured in the conservation stage are set free. Think again of the forest example often used by resilience ecologists (Holling himself came up with the back loop concept while on a walk in the woods[22]). In a post-fire forest, organized carbon and nitrogen, decomposers and producers, feedbacks of sun and water, nutrients and biomass, previously bound up in certain configurations to feed the mature forest, are scattered and released.[23] "Now suddenly," writes Holling, "[is] the time where unexpected events happen. The accumulated resources are disassembled, broken down, left uncontrolled". This is the "reorganization" phase, represented by α, where potential, previously bound up, is freed up for new, unexpected combinations.[24] As illustrated by political scientist Thomas Homer-Dixon, "it's as if somebody threw the forest's remaining plants, animals, nutrients, energy flows, and genetic information into a gigantic mixing bowl and stirred."[25] Space is opened for new species to colonize the area. Pioneer species sprout from stumps of burned trees. Birds nest in their charred branches. Genetic mutations prove useful. Undergrowth is cleared, making way for the floor receive sunlight. Ash settles in, returning previously locked-in nutrients to the soil. Surviving species are freed from long-standing relationships, available and open to new combinations, exploring the new zone using seeds in the soil, debris and existing vegetation—"biotic legacies"[26]— left behind by the disturbance and creating new combinations and feedbacks, testing out new predator-prey relations. The back loop, in short, is a time of great possibility, where the previous forest may be reestablished via existing seedbanks, but novel "unexpected synergies" between invasive and native species may equally give rise to one or many other new arrangements.[27] Resilience ecologists also point out that all back loops are different. Just as possible as the rise of new structures, is the possibility that no new structures may arise.

Instead of a new loop or loops, there could be only phase shift after phase shift, cascading pulsing leading only to dissipation.

Holling summarized these ideas in a heuristic model he called the adaptive cycle. In the now iconic image—it has graced the cover of Holling's recent book *Panarchy* and was even represented in a sculpture—the adaptive cycle is depicted as a horizontal figure-eight, with a front loop of growth (r) and stability (K), and a back loop of release (Ω) and reconfiguration (α).[28] While the idea emerged in Holling's work on insect predation in forests, he and others compared a series of case studies over time—New Brunswick forests, the Columbia River Basin, British Columbia fisheries, Chesapeake Bay's watershed, Austrian alpine villages, south Florida's Everglades—and concluded this cycle could be used to describe the life of each of them.[29] Holling even came to understand his own life through this lens, describing it as following "7–10 year cycles of unplanned intellectual growth, frustration, and renewal."[30] Today the heuristic has been adopted by most resilience thinkers, who bring to it their own uses and emphases, and the heuristic expanded to multi-scalar nested adaptive cycles or panarchies.[31] But across these different emphases the basis concept remains: all systems—human beings, swamps, forests, companies—cycle through a front loop of growth and stability and a back loop of release and reorganization.[32]

The back loop is the least studied aspect of systems.[33] But I would argue it is also the most fecund. While, as I will discuss shortly, resilience proponents generally advocate for the governance of the back loop so as to prevent the loss of a system's identity—to keep systems cycling through the adaptive cycle as in an infinity loop—it is clear that, within each loop's course, there is the possibility for a vast opening of fundamental reorganization or, in a less teleological sense, a period in which new arrangements and possibilities can be worked out and countless divergences launched.[34] In a 2004 paper published in *Ecology and Society*, Holling asked himself whether the adaptive cycle could describe not just regional change, but global and international.

> Are we in a "deep back loop" that presents the same opportunities and crises as the regional back-loop studies that we have described?[35]

Holling's remarks were off the cuff, suggestive and non-empirical, but we can pick up his thread and take it much further, using the adaptive cycle as a lens with which to see the Anthropocene.

Stitching the whole Earth together (Anthropocene front loop)

Within geology, the Anthropocene has generated an impassioned debate concerning its status and chronology, with early efforts dating it to around 1800 with industrialization and the combustion of fossil fuels in England.[36] Likewise, many narratives date with urbanization and proletarianization.[37] Others have proposed that it began in 1610 with the genocide of Native peoples in the Americas while more recently, the "Great Acceleration" has taken precedence, with the Anthropocene Working Group calling for the beginning of the formalization process.[38] Each of these Anthropocene periodizations is important in their own right, and such attempts to measure and demarcate humanity's stratigraphic impact birthed the important study of technofossils, implicating a wide variety of phenomenon, including the Columbian cataclysm, the first atomic bombs, the proliferation of plastiglomerates, and the settling of soot in some of the world's most pristine environments.[39] Yet insofar as these proposed dates seek an origin, asking when it began, how long it may last, and outline appropriate metrics, they do not fully capture the strangeness, disruption and temporal transformation of the Anthropocene as phenomenon. As cultural theorist Daniel Hartley has noted in an insightful essay, "the temporality of the Anthropocene as a periodizing category is bizarre ... shifting as it does between the present, a retroactively posited past and an imagined future."[40] What if this bizarre temporality—the bizarre temporality of our present—is what makes the Anthropocene so powerful both as a conceptual lens and as a historical moment?[41]

To preserve rather than eliminate this strangeness, perhaps the Anthropocene is better thought as having a front loop and back loop.[42]

The front loop refers to the historical periods and processes typically referred to in histories of the Anthropocene and Western liberalism: colonization of the Americas, slavery and economies of resource extraction, as industrialization and the combustion of fossil fuels in England catapulted segments of humanity out of the biological old regime in which humans harvested energy from sunlight, water, plants, wind, and animals, and into a world of factories, proletarianization, and wages, transforming people and environments into resources.[43] Imaginaries of a world split in two, with nature on one side and humans on the other, both seen as resources to be governed and purified, quantified and exploited, as well as the material production of environments in this image and the production of an abstract relation to them, were each moments of the same process of transformation.[44] Dams and bridges, massive in scale and composed of thousands of tons of concrete and tempered steel: such infrastructures,

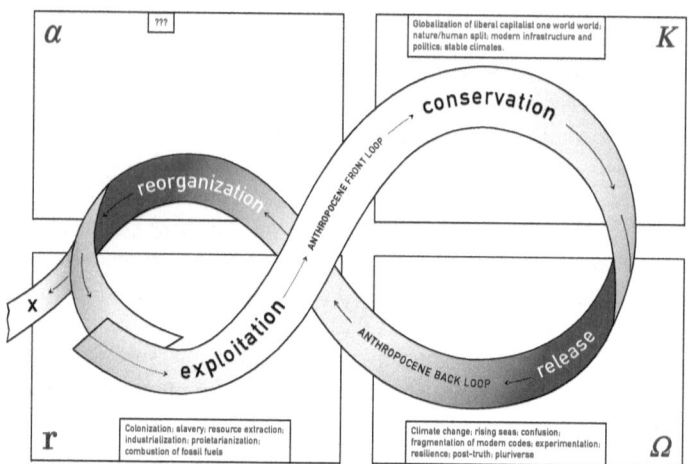

Figure 1.3: Anthropocene adaptive cycle. Design by Caroline Castro for this book.

along with the architects who built them, were in the 19th and early 20th centuries hailed as potent evidence of Western civilization's Promethean ability to shape and order both human life and powerful natural forces, bringing stability and order to cities while transforming natural forces into usable flows.[45] Levees, spillways, and dams were erected to control and regularize the flows and floods of the Mississippi; railways and electrical lines laid out across the United States; all facilitating commerce and urbanization. For city planners behind these audacious engineering feats, among them Robert Moses, "the beauty of public works surpassed that of nature."[46]

Likewise many of the world's peoples were removed from land or forced onto a path of migration toward cities for wage work. Cities, factories, and even schools were reshaped into grids of order to discipline, coordinate, and increase the populations' productive powers and create docile subjects.[47] And even if the mastery and happiness promised was a fiction, it was a fiction that functioned. Looking back at Polaroids of grandparents smiling in front of massive dams or bridges, there is a sense, believable at the time despite most evidence to the contrary, even for working class families, of being a part of an order that was going somewhere better. In the mid-twentieth century coincidence of Fordism and the Great Acceleration, brief windows of possible stability for swathes of the American population opened, built on years of divided and circuitous struggle for industrial democracy and civil rights.[48] With stability of course meaning assembly line production at a rate of every 30 seconds of the rest of one's life.[49] With

Figure 1.4: Family visiting the Grand Coulee Dam, built on the Columbia River in Washington between 1933 and 1942 for irrigation and hydroelectric power. Photo circa 1950.

the building of the (segregated) suburbs and urban ghettos, the workers' movement, with its lumbering union bureaucracies and rank and file masses concentrated in various industries fought for and achieved powerful wage and quality of life gains, as well as a positive, albeit circumscribed, sense of identity.[50]

In this bubble, politics was understood to be the domain in which transformation and liberation occurred. The quest of modern philosophy and politics was to determine being by giving it a name, a ground, or telos. What mattered was some abstract realm beyond or below life that gave it meaning or order. By identifying this safe operating space, outcomes were seemingly guaranteed, or at least stable theoretical pictures of them were possible: justice, equality, a perfectly ordered world where workers would control the means of production, conflict would vanish, and, everything finally equivalent, rivers would flow with lemonade.[51] Thus were the images of liberal life peddled by both capitalist and socialist Western governments and reflected back by those governed. Amidst this a single vision of human life conceived in the terms of liberal governments was forwarded, the Western/Cartesian notion of the human subject—seen as separate from the environment and amenable to governance, variously a nugget of labor power or a docile subject to be shaped and molded by external forces.

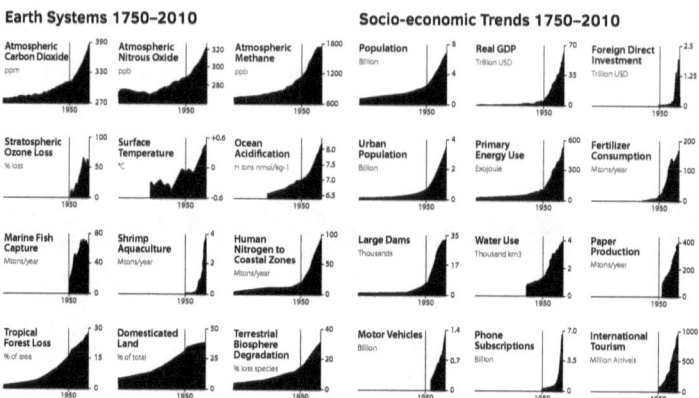

Figure 1.5: The Great Acceleration in socio-economic and Earth system trends. Adapted from Steffen et al.

With reality portrayed as a single, knowable canvas, an ongoing war was waged against other ways of living deemed undesirable.

Together these physical and metaphysical transformations framed a narrative of modernization—of improved standards of living, humankind's triumph over nature, conquering wild landscapes and peoples, productivity and progress—as well as the hegemony of a "one-world world," as John Law has put it, made material pipeline by pipeline, cable by fiber optic cable, powered by power plants and fossil fuels, massive factories housing huge concentrations of wage workers.[52] In the post-WWII period—think here of Earth scientists Steffen et al's Great Acceleration charts (Figure 1.5[53])—international institutions like the United Nations, International Monetary Fund, and World Bank promoted highways, irrigation, and electricity as central to development and socioeconomic advancement, the material substrate for each nation's path forward into integrated global commodity chains, with open trade and flows essential strategies. In the stories of civilization forwarded, the 20th century was progressing toward a single, homogeneous and Western, liberal mode of life based on a unitary standard for thought and action, a "whole Earth" stitched together ever more tightly with global commodity chains, pipelines, and fiber optic cable.[54] Across diverse versions of this imaginary of equivalence what dominated was a vision of progress and order, going toward a future promised to be different and better. The literal ground to this figure of life was the "just right" Holocene interglacial.[55] For the last 11,000 years, the atmosphere trapped just enough of the sun's energy, giving Earth the Holocene's "remarkable long summer"[56]— ice caps at the poles, oceans at just the right pH teeming with life, fresh water rolling from aquifers and watersheds, clean air to breathe, a planet rich in life inside a protective stratosphere. Though often invisible, this was the basis of the Anthropocene front loop's short-lived, but self-described stable and linear world. The safe operating space in which liberal civilization was born, in whose "end of history" we grew up, and which we are now leaving.[57]

Splintering whole Earth (Anthropocene back loop)

It would of course be giving far too much credit to front loop regimes to accept standard narratives of them as truly stable or as a well-thought-out conspiracy to order the world. Various front loop governance techniques were developed to manage, to administer, to respond, to order, in a tangle of engineers, pencil pushers, designers, courtrooms, white papers, petroleum tanks, rebar, and asphalt. They were all intensely ad hoc, local, and reactive, producing disastrous effects along with the much-touted benefits and stability. Today their contradictions have become apparent.

As increased concentrations of carbon dioxide and methane trap more of the sun's heat, Earth's average surface temperature is now rising, with the ocean absorbing the bulk of the warming. Glaciers in Montana and Iceland, but primarily Antarctica and Greenland—formed in past ice ages as water evaporated from the oceans, snow accumulated and compressed, layer built upon layer, under the weight of accumulating seasons, as lower snow became ice, formations became glaciers, and ancient air was preserved, trapped as bubbles within ice—are melting, sending their water into Earth's swelling seas. According to data gathered from microwave and GPS sensors installed on land and ice, in oceans and in space, sea levels have been rising throughout the 20th century and the rate of rise has increased in recent decades, with .1 inch per year in the 1990s to a current approximate rate of .13 inches per year.[58] With this rate increase accelerated by ice melt, the amount of sea level rise projected for 2100 may be doubled from those figures based on a constant rate of rise.[59] Beyond seas, Steffen, Crutzen, and McNeill suggest Earth has left the Anthropocene's first industrializing phase and entered into a second one post-1945, marked by a rising ticker of anthropogenic-induced tipping points crossed or neared: fisheries collapse, biodiversity loss, nitrogen cycles, ocean acidification and coral reef bleaching, as well as deforestation.[60]

In pace with mass extinction, coastal inundation, and drought is also the withering of the certainties to which Western liberal societies tethered their populations, the fracturing of the grounds that gave sense to what could be done and known. Concurrent with post-1945 Anthropocene indicators was the dismantling of the Fordist edifice, which pulled the rug out from beneath the feet of whole populations of workers, canceling the promised future and increasing consumptive musts while making their fulfillment more and more impossible. After decades of revanchism and counterrevolution—wage cuts; deunionization; urban crackdowns; switching from social control by welfare to hyperincarceration; massive wealth gap and soaring profits for the very wealthy—the death of the so-called American dream is felt widely and across diverse sectors of society.[61] Many of the most important social, economic, and political structures which made the 20th century liberal subject—and indeed more broadly liberal ideas of life, politics, and thought themselves—appear possible have been systematically dismantled. For resultant surplus populations, work is defined by low pay, informal, and unsteady work.[62]

The claims to governmental mastery of the world and human life are being washed away by rising seas and unprecedentedly powerful storms—as much as by Twitter feeds—while terminal diagnoses of Western civilization proliferate as quickly as fantasies of the end.[63] Infrastructures once heralded as feats of civilizational mastery are today a key concern not for the glorious order they represent but due to the threats they are seen to pose (cascading network failures, release of greenhouse gasses, toxic waste, targets of terrorism). Global interconnectivity has rendered social political systems not only more chaotic and unpredictable as well as prone to disruption but as analyst John Robb notes has fundamentally changed how they work, with traditional forms of planning and expertise breaking down, nonlinearity leading to destabilizing events, and in other cases complete collapse of legitimacy for traditional institutions.[64] What William S. Lind calls "fourth generation," nonstate warfare predominates, while amidst the crumbling of once-coherent narratives, fractious discourses and new tribalisms of all stripes—religious, clan/gang, made up ties to a made up past or created anew on 4chan and 8chan—abound.[65] In this unique moment, as Peter Sloterdijk notes, "both of the old Anglophone empires have within a short period withdrawn from the universal perspective" and politicians such as Trump "instinctively subvert the norms of modern governance."[66] Amidst the growing absence of unifying, coherent social order or world, social media and tech companies are stepping in in hopes of regulating conduct, to govern and lock in place the data they have made of us. Or it is more accurate to say that human subjectivity is transforming from liberal personhood not only into what Weinstein and Colebrook describe as "a disturbance and a vibration orienting around the chaotic intensities that swirl in the absence of a concept of life as a controllable, containable, nameable force ..." but toward forms of cybernetic immanence?[67]

As Brad Evans and Julien Reid aptly sum it up, "we are living out the final scenes of the liberal nightmare in all its catastrophic permutations."[68] But part and parcel with this are new ways of living and knowing now embraced diversely across place, an expanding universe of trajectories fissuring what was once dreamed to be a single world order. Alongside ever-amassing accounts of colossal earthly transformations that proceed regardless of human involvement—epitomized in modernity's culmination in a "world without us"—equally and together with what are indeed widespread catastrophes, the Anthropocene's back loop is just as importantly defined by a vast proliferation of experimentation in redefining what human life will be. In their own powerful ways, these back loop experiments attune themselves to warmer, wetter, or simply changing worlds, to the upending of deeply held notions and environmental conditions. SpaceX "Mars I" dreams of another space. Pleistocene Park cowboy ecoengineering and dreams of bringing back the past to the present. De-extinction, rewilding, and efforts to recreate the past. In cities worldwide, planners, designers and governments dismissing modern infrastructure as outdated and experimenting with soft, ecological, even "living" infrastructures designed to build resilience. Neighbors and families set up makeshift gyms in backyards and empty lots, experimenting with what bodies can do. We are amidst a wave of experimentation with new ways of transforming bodies, minds, lives, and the world around them: from hacking, making, modding, prepping, and weight lifting to citizen science, eco-design, solar energy grids, and wireless mesh networks. People everywhere are searching their souls, scouring the Earth for tools, and trying in a million ways to reinvent what it means to be human and to dwell on Earth. But together with this search, since 2011 we are also in an era of riots, insurrections, and revolutions from left to right that, to the front loop mind, may look insane, but are very real.[69]

Nature is experimenting too. As global warming has decreased the number of days below freezing, mangroves' habitable range has increased and the trees are taking root in salt marshes farther north.[70] Alligators are adapting to live in residential areas with lakes or canals and use south Florida's waterways as a network of highways to get around. Florida's Everglades are also inhabited by a large population of Burmese pythons—brought to the area as exotic pets and discarded. Despite a state-organized "Python Challenge" that awards cash prizes to freelance citizen groups who catch the most pythons, the release of an iPhone app for crowdsourcing python sightings, and the state's importing of snake-catching specialist Irula tribesmen, the pythons continue to thrive and multiply in their new environment.[71] Seasons are shifting: spring is coming earlier in many places, while winters as we know them in New York have grown more erratic and 60—70 degree temperatures increasingly frequent. "Insects are emerging earlier; birds are nesting earlier; plants are flowering and leafing out earlier."[72] We create our worlds in nature's transforming worlds, and vice versa.

Now what?

In short, if the front loop was the safe operating space of the Anthropocene—here understood not only as a "geo" but also a "geo-social formation" built on a transcendent terra firma of thought and action, however fictional that may have been—this complex, non-linear post-truth world of fragmentation, fracture, dissolution, and transfiguration is what I propose we call the Anthropocene back loop. [73] The back loop is our present, the moment of the naming of the Anthropocene (as a failure), in which the past (front loop) has not disappeared, like points trailing behind on a line, but is erupting in unpredictable ways in the present. However fictional they may have been, the ties that bound—the feedbacks that wove?—the Anthropocene stability domain are coming undone.

The Anthropocene, which literally means Epoch of the Human, has received extensive criticism for its invocation of a single figure of Man (or The Human or Anthropos), which authors have taken to task variously for what is seen as its erasure of race and gender difference or its elision of the fact that the destruction now wrought by "humanity" is in fact caused by the actions of a very small percentage of wealthy humans.[74] While such arguments contain much truth, in my view its invocation of a single definition of human life is the Anthropocene thesis's greatest virtue. For this way of thinking about and molding life within one frame is that of modern liberal regimes, which the Anthropocene thus refers to as an historically specific—dated and finite—strategy for approaching human being. More specifically, the Anthropocene front loop names the liberal project of defining and enforcing life as a one world world in order to call all of this a failure, evidenced in the degradation of natural environments and human subjectivity alike. Thus the Anthropocene back loop provides a name for the liberal way of life as one finds it today: a sinking ship increasingly taking on water from all sides. Instead of heralding the building of a one world world—processes congruent with the front loop's ascendant phase—the moment of naming the

Anthropocene—the back loop—is one of confusion, chaos, and the potential for transformation.

For many this situation can only be perceived catastrophically. Within ecology itself, negative characterizations of the back loop dominate, with the latter depicted primarily as a time to be avoided or governed and illustrated with images of destruction and chaos. Indeed, as I will discuss in the next chapter, ecologists and resilience advocates often seek ways to keep systems in the stable front loop zone, or at least minimize back loop disruptions. For others the back loop is experienced variously as a chaotic time of confusion, fragmenting, crisis, or upending. To paraphrase writer Gretel Ehrlich, is this a world coming apart, or piecing itself back together? Either way, that the old world is finished seems clear to everyone. Whether that is a blessing or a curse depends on one's vantage point.

Viewing the Anthropocene through the adaptive cycle lens, and in particular our threshold now of scrambled grounds, discombobulated modes of knowing and being as a back loop, has a number of benefits. Chief amongst these is the ability to see the Anthropocene not as a tragic End or world of ruins, but a scrambling where possibility is present, old codes are becoming unhelpful, and the future more open than typically imagined. But as will be seen throughout this book, using the back loop to view our time also requires we push resilience thinking's own boundaries, especially as pertains the deep potential for transformation at the heart of its foundational heuristic.

As observed in ecological systems, the back loop is the phase of life in which individual organisms or small groups of individual organisms interact across previously unbridgeable divides and in so doing create something fundamentally original. In contrast to life in the regimes we are leaving behind, where innovation was stifled and influence limited to a few actors with the greatest power—the stability "trap"—in the back loop beings and things are released and open to new potentials.[75] Although most back loops studied by ecologists have been regional in character, in 2004 Holling penned an essay suggesting that "we are at the time of a large-scale back loop," a global situation in which "each of us must become aware that he or she is a participant."[76]

I think Holling's challenge is important; but it is also an apt description of a phenomenon already underway. If we're in the back loop, the question becomes, how to respond? Try desperately to maintain the old safe operating space, freeze a process already in motion? Or let go, allow a time of exploration and experimentation, see what possibilities life holds and what it can become?

2 Government in the Back Loop: Resilience and Managing Safe Operating Space

Governing safe operating space

In the face of back loop dislocations, resilience has emerged as the dominant methodology and discourse under which a host of technologies, designs, and visions are being gathered in hopes of managing urban and global systems in their "safe operating space." Developed by C.S. Holling as a mode of managing the adaptive ecosystems described in his research, resilience is defined as "the capacity of a system to absorb disturbance and still retain its basic function and structure."[77] In contrast to what are now seen as outdated front loop modes of management that sought to maintain a single stability state, resilience is heralded as a form of back loop management that seeks to create and define "safe operating spaces" able to absorb and manage, rather than eliminate, disturbance. Seen as a scientifically verified new worldview, resilience management has risen to the top governmental agendas, from global institutions like the United Nations to city government and activists.[78] As I will argue in this chapter, resilience is a mode of government proper to the upheavals and exigencies of the back loop.[79]

At the global scale, one finds the efforts led by Stockholm Resilience Centre executive director Johan Rockström and host of Earth and social scientists to identify and govern the "planetary boundaries" of the front loop's safe operating space.[80] In a 2010 TED talk indicative of resilience's attitude towards the back loop, Rockström compared our situation, of being close to or beyond the thresholds of the stable Holocene, to a photograph of a man standing at the edge of Victoria Falls, the massive 350-foot-high waterfall in Zambia (see Figure 2.1). "You don't want to stand there!" he warned. "In fact," he continued, "you're not even *allowed* to stand where this gentleman is standing, at the foaming, slippery waters at the threshold. In fact there's a *fence*, upstream of this threshold, beyond which you are in a danger zone."[81] In response to what they perceive as a world on the brink, Rockström and an international team of scientists have proposed the identification of the Holocene's key Earth processes, and management of "a planetary boundary—a fence—within which we have a safe operating space for humanity"[82] (see Figure 2.2.). For Rockström and colleagues, the ultimate goal is global institutional collaboration to manage thresholds and maintain the safe operating space that undergirds "our way of life ... and how we have organized society, technology, and economies around them."[83] This safe space of the front loop is for Rockström the only known Earth system capable of supporting the modern liberal way of life and thus must be preserved. Yet he, along with many other scientists, also agrees that we are already leaving the Holocene and entering the new world of the Anthropocene.

Rockström's response to the back loop, though colorful, is exemplary of the broader spirit of resilience as it is understood at diverse scales, perhaps most ubiquitously in design efforts underway in coastal cities to maintain systems and prevent the crossing of thresholds. In this vein, cities like New York and Miami are now seen as "first responder" laboratories for resiliency infrastructures and strategies for climate change, rising seas and natural disasters.[84] As city governments, designers, and communities search for responses to these new conditions—witnessed recently in Hurricane Harvey's inundation of Houston, the wildfires in the American west, and the devastating power of Hurricane Irma and Maria in the Caribbean—resilience is celebrated as "a vital calculus for any city in this age of uncertainty."[85] "There's no other way," declares a recent film by the Rockefeller Foundation, this is "the resilience age."[86]

Governing urban risk, front to back loop

Cities of course have long been laboratories where shifting modes of governing life are trialed and modulated, and urban resilience as we

Figure 2.1: Photograph of Devil's Pool at Victoria Falls, Zambia, shown by resilience proponent Johan Rockström in "Let the Environment" to illustrate a place "you don't want to stand!" Photograph copyright: Annie Griffith, National Geographic, 2012.

now know it marks the latest configuration of this much longer history of biopolitical government. As put by Michel Foucault, biopolitics—the way in which regimes maintain control over and administer cities and populations via the securing of life—does not proceed primarily as a matter of governments nor preconceived plans, but rather through ad hoc-arrangement of techniques. Such techniques include discourses, practices, architectural forms, regulations, laws, knowledges, technologies and designs, brought together in response to a crisis and that together and in their relations form what he called a *dispositif* or apparatus.[87] Surveillance systems, militarized architecture, urban sanitation, street lighting, policing practices: so many security techniques are responses to other actions and forces: riots, disease, crime, terrorism. Individually and together the techniques of liberal regimes thus proceed reactively and in an ad hoc manner, developing and stitching together disparate knowledge, practices and designs in situ to manage situations perceived as problems or crises on the ground—specifically crises from the perspective of the regimes that seek to maintain their power—within shifting social and political landscapes.[88]

In the face of myriad crises, liberal regimes in the front loop long posited themselves as bulwarks against disorder and their deliverance of security and protection of life in the face of such disorders was central to the kind of life and landscapes they actually helped produce. The so-called stability phase of the Anthropocene's front loop was in fact maintained through constant crisis management of this kind. Indeed, this order played a central role in the creation of ways of life deemed acceptable or productive for liberal regimes in the front loop while extinguishing others deemed unacceptable.[89] As Tim Mitchell recounts, whole cities such as Cairo were reshaped into a grid of order, machines composed of army barracks, schools, and factories whose aim was to discipline, coordinate, and increase the country's "productive powers."[90] Open and orderly streets free from impeded circulation and visibility were a key means of eliminating crime but also constituted the conditions of possibility for the liberal subject and helped shape life in its image.[91] Just as nature's soft, muddy, or meandering environments could be governed with hard, clean, or metal structures of electrical power, dams, and highways, uncooperative populations could be disciplined with pipelines and automated grids, laid out by companies to neutralize worker power to strike.[92]

Since the 1960s, as the front loop probably began to give way to the back loop, apiece with structurally-led disinvestment and ghettoization, the American city has increasingly been construed in the minds of managers, sociologists and governments alike as a place of

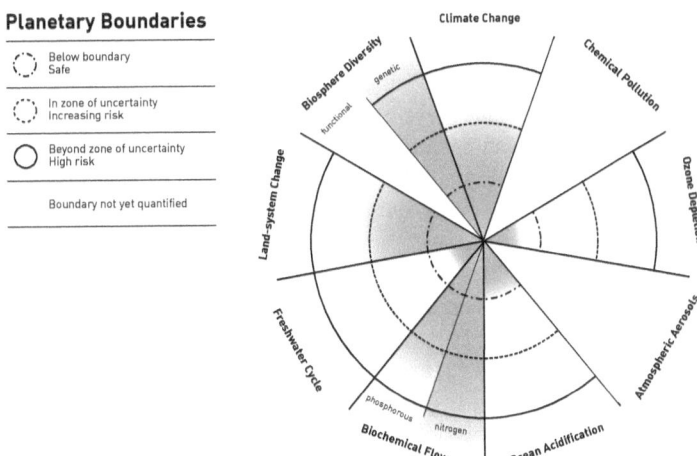

Figure 2.2: Planetary boundaries: a safe operating space for humanity, adapted from Steffen, Will et. al. "Planetary Boundaries: Guiding Human Development on a Changing Planet," *Science* 347, no. 6223 (2015).

crisis under perpetual risk that must be secured. Working through an "ecology of fear"—in which everything from garbage to graffiti, striking workers and panhandling were recast as threats—new forms of management were devised based on embedding security into urban environments.[93] As has been well-described by geographer Neil Smith, this effort was an attempt by the bourgeoisie to reassert control through a sustained warfare on the poor and working classes through militarized redesign of public space, public and private-led gentrification, and embedded surveillance, backed by intensified policing.[94] Instead of trying to normalize or include the city's working class and marginalized populations, in New York this "revanchist" assault brought forth a new sterile urban geared toward commuting, tourism, and investment.[95] Ultimately, past models of liberal social control were augmented by more explicitly militaristic models of governance based on constant securing and ever-intensifying policing of undesired populations.[96]

More recently, with the rise of "networked societies" and the post-September 11 "war on terror," technical systems, like unruly human populations, are seen by governments as posing a source of uncertainty and potential menace to urban orders.[97] In the 1960s companies automated their critical infrastructures, increasing profit rates but also creating systems able to maintain function amidst strikes, thus contributing to more hoped-for order. Today however, interconnected critical infrastructures are now seen as vulnerable to threats coming from ever-expanding list of outside threats—terrorists, hackers, eco-saboteurs, bored kids, and protesters—within broader infrastructurally-comprised environments. But what governments and companies now seek are ways to defend these complex infrastructural systems not only from the outside but from themselves. As during the 2003 northeastern blackout in the US, a very small incident such as a downed power line or faulty switch can now rapidly generate cascade effects across vast distances.[98] Thus for some time now, notes geographer Jon Coaffee, "cities and regions are attempting to embed security and risk management features into their built environments and their systems of governance as part of a drive towards more 'safe' and sustainable communities."[99] Through strengthening existing infrastructural systems and creating new and more systems the drive to "the 'securitization' of network-based urban societies," writes Graham, "becomes such an overpowering obsession that it is used to legitimize a re-engineering of the everyday systems that are purportedly now so exposed to the endless, sourceless, boundless threat."[100] "Everything," continues Graham, "from the design of subways, through the topology of water networks, to the thickness of airplane doors and the software that makes electricity systems work, becomes a site of subtle militarization.... discourses of 'security'... saturate, and militarize, the tiniest details of everyday urban life."[101] Within this field of critical infrastructure protection, infrastructures are now seen as vulnerable to catastrophe, the source of catastrophe, and yet more than ever they are also viewed as carrying the hope of managing catastrophe. Replacing their past status as icons of order, we now hear of hyper-vulnerable and securitized "cities of risk," where discourses of risk interchange with techniques of security, the one folding back into the other, feeding into and augmenting one another in an ever-expanding environment of paranoia and fear, well-illustrated in New York City's new Domain Awareness surveillance system, which integrates over 3,000 closed circuit television cameras around the city with Microsoft software allowing the New York Police Department to crosscheck and monitor criminal databases, measure radiation levels, scan license plates and even human faces.[102]

New York City as lab

Thus has governance of the city stretched to unimaginable dimensions hoping to maintain its own order, as the front loop nevertheless gives way to something less controllable. But climate change adds new dimensions to this story. Amid Hurricane Sandy's wreckage in 2012, for example, architects, designers, planners, and politicians reimagined New York as a fragile city menaced by not only social risks but also ones ecological and infrastructural in nature. A slew of facts now situated the city in space and place—"you know, Manhattan is an *island*"—but not as an ecological matter of getting back to nature, rather one of facing up to water as an unavoidable, threatening presence.[103] Instead of hunter green eco-urbanism, coastlines and neighborhoods were cast in shades of red, orange, and yellow levels of risk and the cityscape marked in varying degrees of vulnerability. These new landscapes referred to households without power, flood areas and casualties caused by storms, but also painted a picture of what is now endlessly promised to be the city's apocalyptic new future. NYC's Office of Emergency Management and FEMA's revised flood maps doubled the number of homes in the high flood risk zone. The *NYT's* interactive guide to sea level rise titled What Could Disappear enabled readers to envision major American coastal areas according to various levels of sea rise.[104] Perhaps most dramatically, *National Geographic* featured a cover story on "Rising Seas" with the Statue of Liberty halfway underwater, while *Nature* magazine's post-storm cover perhaps best summed up the new orientation: "New York vs. the Sea."[105]

For the world's elites, the stakes after all are clear: "we have to do a better job," then-mayor Michael Bloomberg declared after Sandy, "not only keeping our networks up, but keeping our markets and businesses open, come hell or high water."[106] Recent storms, and back loop disruptions more broadly, are generating a climate of existential urgency, setting in motion a widespread search for solutions. For politicians along with designers, planners, and higher education institutions, maintaining business as usual "come hell or high water" is not a matter of just more management needed, but is now cast as requiring a new idea of management itself. In this vein, slews of articles and opinion columns now devote themselves to debunking old approaches of management. City papers welcome readers to the "new normal" *ad nauseam*, calling on readers to abandon dreams of mastery and equilibrium—"the world doesn't work that way," explains resilience pundit Andrew Zolli, calling instead for new ways "to manage in an imbalanced world...in constant disequilibrium."[107] Instead of making the city smart or sustainable they now ask, how to make cities resilient in this new world of omnipresent risk and uncertainty.[108] While sustainability, many experts agree, was an impossible quest "for ways to put the world back in balance," they promote resilience as a more "realistic."[109] "We're not going to make the mistake of fighting the last war," Bloomberg declared post-Sandy.[110] Instead of continuing to block out nature or disorder, resilience is said to welcome such intrusions and entanglements in hopes of managing through attenuation and interconnection. Connected locally and globally, the urban system now imagined is not that of a harmonious, balanced network of interconnection, as in the ecosystems and closed loops of the 1960s, or the blue-green urban nature of sustainable urbanism,[111] but that of an out-of-control, careening landscape of complexity, uncertainty, and risk. "Forget sustainability," cried a *Times* headline post-Sandy, "it's about resilience."[112] City planning commissions, art exhibits and newspapers all repeat similar catchphrases, creating an echo chamber in which resilience rings as the best and only refrain. In place of front loop forms of urban administration, which sought to block out crisis, volatility, or risk, resilience is heralded as a new paradigm that welcomes such intrusions and entanglements, views crisis as inevitable, and, seeing cities as coupled social, ecological, and technical systems, aims to develop their capacity for absorbing or withstanding turbulence.[113]

New realities are said to require new plans, both experimental and audacious.[114] Toward this end, early on New York quickly positioned itself as a test lab for urban resilience. Not only did the city become a "climate change first responder"[115] in a technical sense, but it also rebranded and opened up as a laboratory where new techniques of governing the city in an age of climate change could be tested out.[116] ("New York is where the future comes to audition," has become an often-repeated quote). The Department of Housing and Urban Development put hundreds of millions of dollars into New York City efforts. Meanwhile at diverse other scales ccommissions are constantly organized, expert panels held at places like the New School and SoHo galleries, reports drafted, newspapers' think pieces, with the search for new forms of management forefronted as the key imperative of the future and to "never let a good crisis go to waste."

With the creation of 100 Resilient Cities (100RC), funded with a $100 million investment by Rockefeller Foundation, worldwide 81 cities now have a Chief Resilience Officer, of which Michael Berkowitz, President of 100RC has explained:

> Our goal remains the same—that a mayor wouldn't run her city without a CRO in the same way that she wouldn't run it city without a chief of police.[117]

Once-quaint ecologists from places like Sweden are tapped for expert advice and Dutch "traveling salesmen" flown in from Amsterdam ("'It's like the Dutch East India Company all over again," as scientist Harold Wanless puts it. "They have expertise to sell, and they are pushing it hard'").[118] Across diverse discourses, one thing is emphasized: old models will not work—resilience government must complex; cannot come from top down onto a blank landscape; requires collaboration between diverse actors and communities; must incorporate local environments in which they are implemented—with diverse voices announcing resilience as a revolutionary break from obsolete models: a movement even. As Dr. Raj Shah, President of the Rockefeller Foundation, describes it: "When the Rockefeller Foundation launched 100 Resilient Cities in 2013, we did so with the hope of sparking a global movement to build urban resilience."[119] Cast in terms of innovation, risk-taking, and as a paradigm breaking with old ways and willing to throw out old models, resilience is portrayed as cutting edge and cool—rather than a top down project of urban security and governance—said above all to require experimentation.

Such thinking is endemic to resilience thinking's perspective on the back loop. As resilience-thinking founder Holling himself puts it, when faced with a back loop—which he emphasizes is a situation of the fundamentally unknown—instead of rehashing old models, one shouldn't "try to plan the details, but invent, experiment, and build."[120] Consequently, in a back loop, continues Holling:

> It is essential to do the following: 1. Encourage innovation through a rich variety of experiments and transformative approaches that probe possible directions. It is important to encourage experiments that have a low cost of failure to individuals, the environment, and careers, because many of these experiments will fail. 2. Reduce inhibitions to change, which are common when systems get so locked up. 3. Protect and communicate the accumulated knowledge and experience needed for change. 4. Promote discourse among all parties involved to try to understand where we are going and how to achieve it. 5. Encourage new foundations for renewal that build and sustain the ability of people, economies, and nature to deal with change, and ensure that these new foundations consolidate and expand our understanding of change. 6. Allow sufficient time.[121]

Following resilience's experimental methodology, elite universities and think tanks, celebrity architects and city commissions are converging on cities in a scramble for high profile grants and competition awards, and the privilege to use cities or parts of cities—blocks, sidewalks, neighborhoods, communities—as living laboratories for their designs and projects and varied goals.[122]

The Rebuild by Design competition, co-organized by the Rockefeller Foundation, US Department of Housing and Urban

Development, and the Presidential Hurricane Sandy Rebuilding Task Force in June 2013, has offered such a platform. Billed as an effort to "drive innovation" ("innovation has to break rules") and "to be a model for the rest of the world,"[123] the competition offered $930 million in a search for innovative, immediately applicable and replicable urban resilience measures. Harnessing large scale federal funding for use by local city governments and designers—"those willing to entertain and support unconventional solutions" as one article gushed—"highly unorthodox" design competition. Entries ranged, "from large-scale urban and multi-functional green infrastructure to small-scale distributed flood protection measures and resilient residential structures."[124] Joining forces with hip designers, art institutions, and activists, New York was recast not as an evil oligopoly, but an experimental, innovative, hip, and inclusive site for moving beyond the myopia of federal government inaction and into the urgent 21st century needs.

Redefining infrastructure

Alongside new codes or laws, the focus in Rebuild by Design and resilience more broadly is on design and infrastructure, both as object to protect and a means of protection, with a search for new resilient infrastructures that can better adapt, respond, or absorb risks and transform the urban into a resilient eco-technical-social system of systems able to get hit, reorder, and continue as-is amidst crisis.[125] But the definition of infrastructure itself—traditionally thought in terms of brick and mortar bridges and roads or networked grids—is undergoing a transformation. Here resilience responds to back loop disruptions not only of an environmental but also metaphysical nature. The dualist ontologies separating subject and object that grounded the front loop are seen by resilience practitioners as outdated, and future survival dependent on overcoming them. "We have a new reality and old infrastructures and old systems," Governor Cuomo intoned post-Sandy, adding that the city's electrical utilities seemed like vinyl records in the age of the iPod: "antiquated, 1950s-style institutions that don't serve our current needs."[126] Far from brick and mortar past—highways, power plants, or massive bridges, now seen as brittle, obdurate, out of date—self-organizing and data-sharing human communities are forwarded as "social" or "human" infrastructural systems, the modulation and management of which is now seen as key to resilience. City residents are instructed in preparedness: neighbors with backyard gardens and disaster go-bags are called on to shovel out, clean up, and rebuild, taught to be knowledgeable in first aid and search and rescue—and, as importantly, to be ready to transmit data to city agencies.[127] Residents trained as such are increasingly referred to as what the Federal Emergency Management Association (FEMA) now calls the social infrastructure crucial to resilient cities. This is because, as the Homeland Security agency itself agrees, prepared and connected neighbors add to a city's ability to bounce back, while isolated and helpless citizens subtract from it. Among a host of related emergency preparedness initiatives, the City of New York's Office of Emergency Management (NYCOEM) has launched a campaign called Ready New York, that encourages New Yorkers to see themselves as integral to the city's preparedness and response efforts—"'Re-sil-ient,' 'synonym TOUGH,' see also: New York City," as one City report put it[128]—offering educational videos and guides to a litany of disasters, a "Ready Girl" super heroine (see Figure 2.3), "Choose Your Own Survival Story" "tween" stories, and guide to building a Go-Bag.[129] This expectation is not only after disasters, but extends to "non-disaster" time as well.[130]

Nature too is being asked to become critical infrastructure as well, explicitly incorporated into a larger meshwork of security infrastructures. In perhaps one of the most remarkable designs, New York State is building two miles of artificial oyster reefs in Raritan Bay off the coast of Staten Island, where it is hoped they will act as "living infrastructure" capable of buffering future storm surge and remediating polluted water (see Figure 2.4). Living Breakwaters, as the project is named, is one among six winning designs in the Rebuild by Design competition. In the design pioneered by Kate Orff and led by SCAPE Landscape Architecture firm, oysters are being asked to act alongside other proposed resilient systems to buffer future extreme events and act as a "first line of defense for Manhattan against storms as fierce or fiercer than 2012's Hurricane Sandy."[131] By mobilizing oysters' natural processes (attaching themselves to each other and developing reefs that adapt to changing sea levels), the hope is to cancel out other natural processes (i.e. hurricanes and storm surges). Titled the Tottenville Reach, the reefs are being built on a pilot site, and understood by designers as a large-scale experiment in real time, providing both opportunity for different firms to engage in new live trial methods devised for the back loop's new exigencies and moreover to see whether or not the project even works.

In contrast to traditional restoration, the living breakwaters are the first experiment utilizing reefs for their "protective function," and are therefore being designed along a different, more biopolitical criteria: they need to function, be efficient, maximize capacity, be sturdy, and able to confront "aggressive" waves. As a result, the construction site is to be situated further off coast, directly in the midst of

Figure 2.3: Ready Girl, super heroine created by New York City Office of Emergency Management, teaches school children about emergency preparedness. Copyright: NYC Emergency Management.

dangerous high wave climates. Just as the resources and the landscape of New York as resilience lab gave SCAPE an opportunity to test new oyster recruitment structures in real time environments, likewise for companies like SeArc—a Tel Aviv-based coastal engineering/consulting firm, who designed one meter by one meter, one-ton blocks or "ecological armoring units" for the project—it is a chance to test their techniques in real time—with the aquatic architecture a pathway into a "living laboratory."[132] While testing has been done in the Mediterranean Sea, Key West, and Haifa laboratories—making the Tottenville Reach its first live trial. Ultimately, it is the hope of designers that oysters will grow on each other, layering onto and strengthening the assemblage to which they are attached: "Designed as living systems," SCAPE's report to the Rebuild by Design jury explains, "they build up biogenically in parallel with future sea level rise," "rising elegantly with the seas."[133]

Further south in Miami

Meanwhile early models test run in Sandy's wake are now being replicated in other coastal cities, with the Bay Area recently launching its own Rebuild by Design competition and the launching of other resilient cities networks. Further south in Miami, Florida, early onset of sea level rise has likewise led to adoption of the resilience modality.

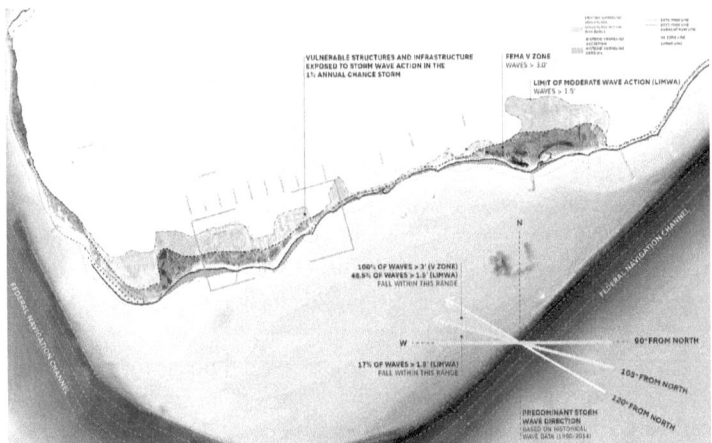

Figure 2.4: Living Breakwaters design for south Staten Island, New York. Image by SCAPE Landscape Architecture team for the Governor's Office of Storm Recovery.

Built as real estate reverie and known for its art deco fantasy image, Miami—crisscrossed by unique, often conflicting metabolisms and trajectories including a hyper-segregated real estate development driven by foreign capital, environmental policies oriented around maximizing land value, and diverse local cultures—is increasingly seen as "ground zero" for sea level rise amongst US cities.[134] Due to its position as one of the lowest-lying coastal cities in the US and foundation of porous limestone, sea level rise is a present-day reality in the city, with sunny day flooding ocean water coming up through the city's sewer system and sidewalks even on non-rainy days six times per year.[135] Current flooding is now captured through iconic images like that of an octopi floating in a parking garage or Miamians wading through flooded intersections on their way to work, and is projected to increase to 80 times per year by 2030, and 380 times a year across a much-expanded territory by 2045, when sea levels are projected to rise 15 inches.[136]

While former governor of Florida Rick Scott publicly denied climate change—and issued an unofficial ban on use of phrases like "climate change," "global warming," and "sea level rise" in government documents[137]—in recent years the City of Miami Beach has taken up the task of building climate resilience. Now labelled a living laboratory for urban resilience,[138] Miami Beach has taken up the resilience mantle in diverse ways including joining Rockefeller Foundation 100 Resilient Cities as part of the Greater Miami and the Beaches collaboration, participating in Columbia University's Center for Resilient

Cities and Landscapes-100 Resilient Cities' Resilience Accelerator program, and working with local organization ArtCenter/South Florida to bring its first embedded artist to help the city brand its resiliency strategies more sensitively using tools of the avant guard.[139]

University think tanks and resilience networks have converged upon the city in the back loop, pouring billions of dollars into the search for new modes of administration. Resilience studies and institutes, expert consultants and funding agencies—Columbia University, Bjarke Ingels Group, Rockefeller Foundation, to name a few—have flocked to the city. Alongside its branding efforts, Miami Beach is currently the site of a projected $600 million, ten-year climate resiliency infrastructure program laid out under former mayor Philip Levine, a "cruise line media magnate" elected in 2013 on a campaign promising to build resiliency to current and future floods, even filming one of his campaign videos leaving work in a kayak.[140] Under Levine, via no-bid emergency contracts, doubled stormwater fees, and drastically multiplied sea-rise projections, the city of Miami Beach fast-tracked a suite of infrastructure projects including installing a fleet of industrial pumps, elevating streets, and building new seawalls designed to prepare Miami Beach for the next 30-50 years of sea level rise, and specifically for coastal flooding, King Tides, heavy rain and storm-related flooding, and groundwater flooding.[141]

As part of the plan, referred to as "Miami Beach Rising Above," in Sunset Harbor, a neighborhood of Miami Beach, engineers have elevated several blocks of road adjacent to a cluster of condos and restaurants. Today walking through the neighborhood's cluster of towering condos you may overhear wealthy condo dwellers debate whether to order the $25 vegan noodle paella with corn alioli or the $29 grilled octopus with fingerling potato foam which is not unusual for the area except for the fact that diners are now seated two and a half feet below the recently raised street (see Figure 2.5). To accommodate the elevated streets, modifications to existing businesses have been made. Publix grocery store chain, for instance, built to what several years ago was seen as base flood elevation, with seven stairs leading from the previous street height into the store's entrance, has now had five of the previously existing stairs leading into the large store eliminated, with just two stairs now leading into the store. Complimenting the raised streets, to build resilience to the regular flooding, Miami Beach has torn up streets to install twenty industrial pumps—costing $2 to $3 billion each—with plans to increase this number to 60-80 in coming years to funnel flood water out of city streets and into adjacent Biscayne Bay.[142] Finally, capping off the Rising Above plan, the city is installing a series of new sea walls to protect streets and property from flooding coming from the Bay.

But with sea level rise, many speculate that Miami's most urgent climate change threat is not flooding or land loss but destruction of the city's freshwater supply via seawater infiltration into Biscayne Aquifer.[143] As such, the once-restorative goals of a U.S. congress-approved new Comprehensive Everglades Restoration Plan—a 30-year, $7.8 billion effort of the South Florida Water Management District and the Army Corps of Engineers to restore historic surface-water flows of the 9 million acre Everglades system[144]—are now reconceived as a nature-as-infrastructure effort to render the city resilient by increasing natural water flows to maintain freshwater pressure against the intruding salt water. Here making the city resilient is understood as a matter of restoring "nature's" flows outside the city, by reversing the U.S. Army Corps of Engineers's massive hydraulic infrastructural control system built in the mid-twentieth century to channel and contain the Everglades' flows. Part of the project is being conducted via experimental techniques in a vastly altered landscape replete with technological and industrial legacies and diverse human communities at Deering Estate[145]–an active experimental restoration site in Miami designed as a trial run of methods to be used in the Everglades proper, a kind of niche space for future "patchy" restoration strategies. Most recently in Miami under the umbrella of the Florida Disaster Resilience Initiative a coalition of politicians, consultants, and academics came together for a day-long live action role playing game—the Serious Games—to imagine and anticipate a future disaster scenario in all its cinematic details, and guide community participants toward the prepared solution of building resilience hubs. Different from New York, all of this is the Miami iteration of climate resilience: maintaining a (static yet ever-receding) pastel and neon art deco fantasy while the oceans rise around the city, with new infrastructures like these streets alongside a delirious post-great Recession building boom, with 1.6 million square feet of office and 1.8 of retail space under construction in the second quarter of 2016 alone.[146] The future is sure to be more resilient, with the recently approved $400 million bond to build sea level rise and flood prevention infrastructure across Miami, part of Mayor Tomás Regalado's vow to fight the already-occurring sunny day flooding: "we cannot allow this to become the new normal."[147]

Figure 2.5: Dining beneath the city's newly-elevated streets, Miami Beach, Florida. 2017. Photo by author.

Resilience's relationship to the back loop

Though resilience recognizes deep-seated problems, its key tenet and goal is to attenuate and govern disruption in order to maintain the identity of the system. It is therefore possible to read resilience in its global and urban manifestations as representing one possible orientation to the back loop, Johan Rockström of the Stockholm Resilience Centre's "you don't want to stand here!" at the foaming, slippery waters at the threshold. Indeed, most resilience literature portrays the back loop in the negative. Illustrations of the back loop conjure images of disturbed, post-catastrophe landscapes, torn apart by forest fire devastation or societal collapse. In other cases, the back loop is the other side to the deadly tipping points that threaten human civilization (the anarchy which the loss of liberal order threatens to deliver). Echoing Rockström's reaction, Walker and Salt write, "given how unpleasant release and renewal can be, it's comforting to know that most systems spend most of their time in the fore loop, which is generally slow compared with the back loop."[148] Likewise, in most accounts movement into a back loop is caused by a crisis event: wildfire, hurricane, or financial crisis, sending systems into the back loop "freefall."[149] Like Rockström, who seeks to ward off the Anthropocene and welcomes us to it in the same breath, Holling shifts between a coming and an already-present back loop, with the world both "on the brink" and "in" a big back loop.[150] In all cases, resilience practice follows with its techniques to "navigate" or safely "pass through" such times, the latter portrayed as something to endure, a crisis to prepare for, a disaster to stay afloat during, bounce back from, or pass through.[151] Ultimately, resilience's goal as a mode of management is to maintain the overall cycle or system, while allowing for its modulation or mutation.

As resilience ecologists see it, endogenous and exogenous disruptive events such as fires, floods, or droughts "test" systems in the stability phase, leading to the uncertain time of a back loop, in which institutions must respond.[152] Management responses may take the form of avoidance, or of efforts to stay in the front loop and maintain stable parameters and approaches. What resilience practitioners recommend instead is a period of experimentation, which embraces the period of uncertainty and uses it as an opportunity to test out new management responses.[153] Such an approach begins from acknowledgement that there is a buffer zone within which regimes can exist while still maintaining their state or identity, and experiments seek to find those thresholds, while testing out new ways to maintain systems. In its current urban resilience iteration, what we might call an emergent paradigm of "back loop urbanism," represents the taking up of this approach to urban systems faced with Anthropocene dislocations such as rising seas and flooding. Rather than deny the reality of these events, and rather than attempt to use past management forms to administer them, back loop urbanism represents the range of experiments underway to adapt and remodulate cities in tune with such transformations.

For some, such resilience experimentation promises salvation: to heal the nature/culture divide by uncovering nature's potential or reconnecting urban dwellers with nature, each other, or Earth systems[154]—as if imagining a fully interconnected city could somehow soothe the latter's volatility. In ecology and systems thinking, fields from which the concept itself emerged and where it is today much discussed, resilience is promoted as a corrective to outdated "stability" models with their notion of ecosystems as single steady state and equilibrium and sought to manage them by preventing disruption or change.[155] For resilience thinkers such as Carl Folke, this is now an "old" perspective, that "provide[s] little insight"[156] and is in fact an "alienated" "pathology" actively preventing progress and increasing vulnerability. In its place, resilience posits what is seen as a more realistic, accurate, and contemporary view of the world as made up of dynamic, fundamentally unstable, coupled, nonequilibrium social-ecological systems.[157] One must now, as Folke puts it, "learn to manage by change rather than simply react to it... uncertainty and surprise is part of the game and you need to be prepared for it and learn to live with it."[158] Though its definition and approaches to its management are seen as works-in-progress, amongst ecologists resilience is seen as having "universal applicability" and is an "inherent property of systems" that today must be developed.[159] Although it was simply a systems error while modelling predator prey competition in the 1970s that gave Holling the idea of resilience, and about which he initially said "I couldn't decide if it was real or bullshit,"[160] today many thinkers believe resilience represents the endpoint of a story of Enlightenment, where resilience is portrayed as the discovery of a salvational "correct" view of the world and of nature. "We *were* Cartesian reductionists, thinking of the world as separate objects," this narrative repents, "but now we're not! At last, we are living in a system!"

However, salvation is precisely what resilience infrastructures tell us is impossible.[161] Rather than promising the future understood in terms of progress and improvement, resilience infrastructure turns the temporality and politics of front loop infrastructure on its head. Instead it assumes a future that is only getting worse, inevitable crisis, and seeks only to minimize their effects, adapting to changing conditions so as to keep existing socio-economic conditions of liberal life the same (or more accurately, on life support). From Rockström to Bloomberg, resilience's power brokers have not been shy about making such priorities public. As Rockefeller Foundation president and resilience advocate Judith Rodin succinctly summarized it:

> This isn't just climate-related. We're not just thinking about hurricanes or floods; we're really thinking about any vulnerability to the system that could take it down, and how to build against that.[162]

What's more, if there is a difference between resilience infrastructures and modern infrastructures, with the former seen as overcoming the problems of the latter, there is also a relationship between the two. Resilient infrastructures are designed to manage destruction and disaster caused by the ongoing function of front loop infrastructures like pipelines, mines, and power plants. Resilience designs thus work *in conjunction* with front loop infrastructures, which continue to produce the disasters resilience intervenes on. Put otherwise, the role of resilient infrastructures is to manage and adapt to changing conditions of catastrophe at sea—rising seas and storm surge—to secure and manage an unchanging urban order on land. Despite their "green" characteristics—oysters, swales, reefs, and marshes—they are deployed alongside the proliferation of pipes, cables, wires and roads that underwrite modern liberal life. Far from a new ecological savvy, resilience infrastructure must be understood as the substrate of a liberal regime promising neither redemption nor progress but only survival of existing, ruinous conditions amidst catastrophe.[163]

Take, for example, Raritan Bay, the site of New York State's oyster reef design, which is crisscrossed by some of the most important shipping routes in the US Northeast and traversed by oil tankers to and from massive petrochemical facilities on the New Jersey side of the bay. According to designer Orff, approval for the oyster test site was contingent on SCAPE's ability to prove that the project will have "no effect" on these ACE infrastructures. A powerful reminder of the way in which infrastructures to manage the effects of climate change literally sit side-by-side technologies that continue to produce those effects in ever greater proportions. Pipelines like Keystone XL or drilling in the Bakkan Shale are necessary to liberal life as it is now lived as well as to its extension into the future, while resilience projects are being built to secure that way of life amidst the very disasters it itself generates.

Resilience's status quo goals can be seen in other ways in Miami. Whereas in cities such as New York resilience projects are often presented as a dynamic and revolutionary new design approach based in complex, entangled socio-technical-ecological systems, across Miami resilience is more typically discussed in an explicitly economic register geared toward maintaining real estate markets, tourism, and investment in the South Florida region. Resilience power brokers such as Levine have not been shy about making the priorities behind their mobilization of resilience public. For Levine and others, the main—and explicitly stated—goal of Miami Beach's resilience infrastructures is to maintain the city's real estate market.[164] What is protected, politicians hope, is a way of life that moves locally between high end developments, yachts and restaurants as well as internationally through off-shore banking institutions, second and third homes, and art fairs in distant cities. Despite being portrayed as vulnerable frontline experimenters, politicians behind such projects have rammed them through via apocalyptic discourse that states there is no alternative, no-bid contracts, and high-powered funding. In a kind of blackmail, future survival writ large is conflated with the survival of this particular vision of the city, an ethos aptly summed up in the name of Miami's new $400 million resiliency infrastructure bond: "Miami Forever."

While it is celebrated by many as the city's or planet's salvation—or final enlightened nature/human relation—what such infrastructures make clear is that resilience in fact forwards a disabling fiction whereby human survival in the Anthropocene back loop is tethered to the maintenance of existing economic, social, and political relations.[165] Indeed the back loop is treated as a time for remodulating governance of these very relations. Within these, resilience's goal is to maintain the single, biopolitical vision of human life forwarded by liberal security, and to maintain it within the one world world that has been installed. On the brink of regime shifts (of truth, water, land, or humanity) resilience qua governmental apparatus can only offer techniques to manage old parameters. Oyster reefs to modulate and manage oceans moving into new regimes, "roadmaps" to reverse Earth systems away from the brink, while underneath communicating the futility of the task: when asked to imagine the longer-term future of Staten Island, oystertecture designer Kate Orff replied, "towers, walls, and jellyfish."[166] Similarly, while Rockström's 2010 planetary boundaries TED talk held the Great Barrier reef up as an example of management success pulling systems back from the brink, today the reef is used to represent another system passing the threshold, beyond saving except for its ecological functions useful to humans.[167]

Implicitly or explicitly, resilience designs therefore tell us that there is little left. In Miami Beach's case, Bvlgari bags will float on the rising seas as they churn outside the city's preserved art deco hotels: a frozen, yet ever-receding past of luxury, glamour, exuberant faith in social and technological progress. To paraphrase Harmony Korine's James Franco, Miami Beach forever... Miami Beach forever...Miami Beach forever. Or in New York, oyster reefs rise elegantly with the seas. The aerial view of a resilient Manhattan is one fringed in peaceful green, but zoom in and we find REM Koolhaas design firm OMA's vision of a new normal day lounging in Central Park, now with airdrops of Flood Alerts and Flood zone maps.[168] Or take, for example, Bjarke Ingels Group's $540 million "BIG U" sea wall now under construction around lower Manhattan.[169] The "Reverse Aquarium," the project's signature building, was imagined as a museum located on the ground floor of the project's flood protection structure, featuring a glass window looking out onto New York's waters, where museum goers watch sea level rise in real time from behind plated glass (see Figure 2.6). Populations worldwide are held in suspense, fear, anxiety, buried in debt and bills and dreams of the end. The possibility of imagining or creating other worlds disappears.

Rather than promising the future—any future at all—as a mode of administering cities in the back loop resilience functions to *ward the future off*. Rather than promising to change life, functioning to ward other ways of living off. Adapting to changing conditions so as to

Figure 2.6: Mockup of "The Reverse Aquarium," signature building for Bjarke Ingels Group's BIG U sea wall for Manhattan, as imagined in original proposal for Rebuild by Design competition. Copyright The BIG Team/Rebuild by Design.

keep all other things the same. In place of what is now seen by resilience proponents as the front loop's wrongly hubristic human, resilience forwards a degraded, anxious subject that must endure crisis as condition for existence, a victim or hostage whose only agency is as stewards to Earth processes, conceived either as life support systems or living infrastructure (or as sensing nodes in a broader "data-intensive and extensive coastal site"[170]). Here crisis-managing infrastructure is the dominant rubric through which all spheres of urban life are increasingly seen. As in Liu Cixin's *The Remembrance of Earth's Past* trilogy, every minute of every day, every second of life lived, becomes a preparation for the end. With the world around us perceived as a constant source of risk, anything that can lessen that risk becomes acceptable. Safety becomes the end all and be all of existence. Climate governance is militarized, borrowing on techniques first developed in battlefield command strategies.[171] Art, culture, the basic details of life become preparation for impending disasters seen as omnipresent and impossible to prevent. Many excellent studies of resilience have been written, but perhaps most evocative of the will to survival at the heart of resilient modes of administering urban environments can be found in fantasy writer Steven Erikson's *Malazan Book of the Fallen* series. In the books, immortality is a curse. A people called the Imass undertake a collective ritual to live forever in order to wage eternal war on their enemies (who, the reader later discovers, in fact lived with a vitality and free spirit unbeknownst to any other peoples). Thousands of years after the ritual, the Imass still wander the continents, haggard, skeletons with pieces of skin falling off, dreaming of nothing, contemplating their own futility and the degradation of their own way of life, remembering little of what it is to actually live: love, sex, tastes, everyday commonplaces like the smell of cooking food, children laughing, leaves rustling, birth and death.

Experimentation and power

While resilience teaches that there's no more dreaming, no more hubris, many of its animators are ironically embracing the very ethos it tells us to reject. In fact, studying resilience one cannot but be astonished at how much actual human agency—powers of imagination, hubris, and tools for their translation into reality—is currently held in the hands of those who want to preserve and profit from the present social system as the water rises. Faced with a society in the back loop—understanding we are leaving Western civilization's safe operating space—the powerful are experimenting and getting organized to thrive. Their experiments may be governmental or even malevolent, but they are also daring, often so much so that they believe they can transform the very cities we live in and the waterways that surround us into large-scale laboratories for their trials. They are maniacally trying to make a future in the images of their desire. What about the rest of us, for whom the safe operating space of existing society is not safe at all? How to move forward? How to break out of this cycle?

If cities are *already* entering the Anthropocene back loop, then why not *embrace* the back loop's destabilization and relational upheaval to open up transformative potentials and other futures? Instead of continuing on the same, eternally, for no reason, why not take this opportunity to reinvent life and make it more worth living?

Despite its dominant iteration as liberal governance, resilience's design modalities might actually provide a vector for thinking such profound kinds of transformation in the back loop. Resilient urbanism is fundamentally a shift toward situated experimentation in which the "lab" is not a pristine or sterile space but rather the urban environment itself, an expanded field in which old boundaries and limits are removed and innovation, imagination, and daring are not only considered essential tools for design but also for urban survival. Yet because it is situated in the world, in local environments, histories, and metabolisms, resilience is equally reliant and shaped by local legacies. As such resilience practices offer a critical method for the back loop: they make use of the world—more on "use" in chapter 7—but also occur in a world that exceeds us and which we do not control. Many of the very same resilience projects, which operate to dim down our imagination, are at the same time themselves deeply imaginative. They are ready to transvaluate—to let go of old models seen as no longer useful. They are experimental yet audacious, speculative and creative, and act in the here and now. Each of these experiments has an in-situ, ad hoc, character. They respond to local conditions, to manage specific local needs.

In keeping with their experimental nature, the efficacy and future of resilience designs is uncertain. Where urban government by experiment past had an uncertain character, today this uncertainty is taking on new dimensions and profundity. Raised streets and pumps in Miami are considered by many a debacle, the former causing flooding for area restaurants, the latter found to be discharging human fecal matter into Biscayne Bay. Oysters in New York may not ever take hold. No can say if it will work. "What we're looking for," Kate Orff has explained, "is a spark, a critical mass enough to jumpstart life again in a place that is practically speaking dead."[172]

As the consequences of more storms, floods, and fires unfold, and as the inevitable tidal wave of think pieces on how to "bounce back" wash in, the post-Sandy stitching together of a resilience doctrine bears consideration. Katrinas, Sandys, Harveys, and Marias: these are events that change us. They scramble our view of the world, shift the ground under our feet, and open up a number of possible trajectories and possible responses. There is always another choice, and while its discourse primarily provides darkness, paradoxically resilience practice may have more to offer. Its grandly ambitious plans and their deeply experimental ethos as well as the relative speed with which its proponents have succeeded in producing a paradigm shift are all something to be admired about resilience.[173] As we search for other paths forward in the back loop, we are each free to use the best lessons of resilience practice and discourse to challenge its worst aspects, to open up a much wider field of possibility and with a much broader set of actors. All of us, in all our various worlds, with our aptitudes, and skills are free to take back the powers of audacity, experimentation, and transformation, and put them toward other paths, alternative meanings for resilience, and more livable ways of life. Up for grabs in the back loop is not only the survival of human life, but its very meaning. It's everyone's right to take up such a question.

3 Sit Down, Be Humble: Imaginaries of Post-Apocalyptic Survival Amidst interlinked Ruins

Imagining life beyond safe operating space

Red skies hang over California, floodwaters lap at America's Gulf Coast, and just about everything else seems to be going down the tubes. While resilience seeks to ward off the end, as we tip into the Anthropocene's back loop—who hasn't wondered if the end has already come?

If this is the case—if instead of awaiting a cataclysmic event in the future, we are already living in the post-apocalyptic—what possibilities for life beyond thresholds? Just like the expanding field of resilience thinking and design, recent film, theory, and literature offer ideas and imaginaries of what life beyond safe operating space might look like. Whereas resilience's imaginaries must be teased out from within its seemingly technical infrastructures, works discussed in this chapter seek explicitly to forward new imaginaries. Although widely celebrated, this chapter argues that these imaginaries require more critical investigation. What practices and forms of life are being forwarded in them? Despite their ubiquity and claims that the Anthropocene "changes everything," are we actually seeing something new in them? Or might we instead be witness to the assembling of another apparatus for governing the Anthropocene's back loop?

To begin let us consider recent film *Blade Runner 2049*.

Set 30 years after its predecessor, *Blade Runner 2049* depicts a future we can really believe in. Ecological collapse has caused widespread famine, the Blackout has wiped away data and thus history, and the only living things are stunted humans, bioengineered replicants, and giant beetle larvae on distant protein farms. Amid the ruins stands the Singularity incarnate: savior-tech-billionaire-freak Wallace, played by Jared Leto, who is based—like the late Tyrell Corporation—in a massive brutalist temple bathed in noir lighting and updated for the era of climate change with the undulating shadows of rippling watery light.

Wallace plays a katechontic role: manufacturing newer, more obedient replicants; controlling the entire food supply; and expanding humanity's reach to nine planets. Policing the existential line between human and not-so-human is our hero Detective K, played by Ryan Gosling, a dreary wage slave of the LAPD. A blade runner, K is tasked with hunting rogue replicants. The movie's opening scene shows him on the hunt, reluctantly retiring an old model unit with a powerful secret. K himself poses a threat, and after missions he undergoes a Post-Trauma Baseline Test to detect emotional response and emergent autonomy—cells interlinked within cells interlinked within one stem and dreadfully distinct against the dark, a tall white fountain played.

Only errant memories, his live-in holographic OS girlfriend Joi (Ana de Armas), and Ryan Gosling's pensiveness seem to indicate his life is anything but meaningless. But when K uncovers a miracle and begins to suspect himself of being a kind of skinjob messiah whose existence could "break the world," love blossoms and fantasies emerge. K and Joi overcome their given conditions—replicant, limited AI—and by dreaming, exploring, and some added techne, create another reality beyond ruins where colors are richer and beings more defined. But is it all part of the program—Wallace's designs, false memories, coding? Is it real?

...

While resilience seeks to hold on to safe operating space—thus reducing life to warding of the end in perpetuity—many in the Earth sciences including Johan Rockström himself at times, state that we have already crossed many boundaries, and that the Earth is *already* leaving the old operating space behind. According to scientists like Owen Gaffney and Will Steffen, the planet is shifting out of the stable climates of the 11,000-year-long Holocene interglacial in which modern civilizations developed (and also out of the glacial-interglacial cycle in which it flickered for the last 100,000 years) and moving on toward volatile and unknown operating spaces, one of which may be a so-called Hothouse Earth,[174] or alternatively the environment of *Blade Runner 2049*. Fueled quasilinearly by carbon dioxide emissions and biosphere degradation, this trajectory is coupled with nonlinear biogeophysical feedbacks creating tipping cascades—permafrost disappearance, land and ocean carbon sinks weakening, polar ice sheet melting—accelerating global warming as well as pathway irreversibility. To reiterate, this shift pertains equally to the geosocial formation built on the terra firma of liberal front loop thought and action as much as the stable Holocene Interglacial.

Safe operating space is unraveling as we enter the back loop, an unknown place of chaotic fragmentation and freefall, but also experimentation and potential, where beings and things are opened to new possibilities. As with every ending of a civilization—there have of course been many—equally widespread are dreams of, and experiments in, new ways of living. Unfortunately, much of contemporary politics and culture is dedicated to holding this deluge back, to manning the wall—disciplining the back loop—beating the lesson into our heads that ruins and bare life are our only hope, asserting in perpetuity, "You know there's nothing else, right?" Infrastructures, design, and actual walls such as those discussed in the last chapter are certainly a part of this, but perhaps most powerful of all are the lessons exercised on our interiors, imaginations, and dreams.

So, back to Joi and K in *Blade Runner 2049*. Was it real?

[Spoiler Alert]

Of course not.

Joi gets free from her mainframe, only to be crushed under the shining boot of Luv, Wallace's "best" replicant. Far from a miracle or a real someone, K learns he is just a decoy to hide the true messiah. His memories were implanted replications, and alone in the night he encounters a holographic ad for the Joi AI™, who beckons, "you look like a good Joe, let me tell you everything you want to hear." Joi is a product, an app just doing its job. K and Joi's love, their truth, and the possibility of another life beyond 2D ruins woven from the two: none of it was real. Joe—the regular Joe like all of us, working away all our days, feeling lost, wondering if we can get a foothold, maybe daring to dream of becoming someone better, somewhere different—is just a nobody, a naïve skinjob who believes in a phantom love, a tool for the revolutionary replicants, a dog for the LAPD, a foil for humanism.

Sovereign grounds and hatred of the self

> "Do you dream about being interlinked?
> Have they left a place for you where you can dream? Interlinked."

The baseline check is actually a poem from Nabokov's *Pale Fire*, in which a fictional poet, John Shade, has a near-death experience where he sees a white fountain and discovers a newspaper article of a woman who had the same near-death vision. Previously a broken man, discovering this incredible connection gives Shade wonder and purpose. When he tracks the woman down, he finds out it was a typo: she actually saw a mountain, not a fountain. The lesson here? Be more realistic; be more humble; stay servile, K.

A brief foray into one of the front loop's key foundations, Christianity, helps to explain this situation. Historically one of the biggest challenges to pastoral government come from the tensions Christianity sets up, wherein the Kingdom of Heaven is to be found in the sky or perhaps inside of you, but never around you. Denigration of self and the profane world was a solution to the endless problem—persistent in the front loop—of what to do with those who didn't want to wait for God (or established church ministers) to show up again: Essenes, living without property or money in the Judean desert; the desert fathers with their Nitrian gardens who found perfection in solitude; the post-crucifixion primitive church with its agape feasts and image of Jesus not as a martyred or ethereal figure but as a dancing sorcerer, a sun; or the myriad millenarian uprisings in medieval Europe... After all, you can't have ordinary people just going around calling the Kingdom of Heaven into existence. It is to be found in the sky but never here; in the future but never now. The role of the *katechon*—the church and empire—with its moral codes of forbidden and permissible conduct; prohibitions and constraints; fear of God and austerity—was also to prevent populations from getting the idea that perhaps kingdom *was* here, now.[175] Or that it could be made by their own hands, spoken with their own lips.

In the more secular present this wall is maintained in diverse ways via politics. In his *Western Illusion of Human Nature*, anthropologist Marshall Sahlins clarifies how politics imagines a world split in two.[176] In one sphere is life, and in the other, the forms or answers seen as proper to the government of that life. In this dualist framework—which erases the fact that throughout human existence people were capable of deciding for themselves how to live and organizing their worlds accordingly—life is portrayed as being in *need* of being governed. As such the world is posed as a problem to be managed, a problem that is then answered through one form of government or another, which is itself set up as an exterior plane of existence. Following this approach, politics gives beings a specific function or *telos*. This, Sahlins writes, is true as much for totalitarianism as democracy or anarchism, in so far as each posit a plane separate from life itself, which will then govern or manage being.[177] Such, he argues, is the singular metaphysics across all politics—right, left, or otherwise—and running deep into our own hearts and minds. If the first lesson we're taught is to not, whatever you do, begin from yourself, and, most importantly, do not trust your real (it's "just" a dream), then the logical corollary is that the real must come from elsewhere.

Expanding this to the broadest scope, the quest of philosophy and politics in the front loop overall was to determine being by giving it a name, a ground, or telos. In place of trust in one's own intuitions, actions, or definitions of truth, codes of sovereign grounds are created, to which being and action will be required to refer themselves. "Doctrines of an ultimate ground,"[178] as philosopher Reiner Schürmann called them, delimited the conditions of possibility within which things can be said, lived, and dreamed, providing yardsticks to which thought and action should be referred and the safe operating space in which they were possible. These grounds—God, rationality, liberal Humanism, etc.—would legitimate ideas of reality and truth, as well as give legitimacy to existing authorities, such as the Church, government, or rulers.[179] What mattered was always some abstract realm beyond or below life that gave it meaning or order. By identifying this safe operating space, outcomes were seemingly guaranteed, or at least stable theoretical pictures of them were possible: justice, equality, a perfectly ordered world in which the rivers would flow with lemonade.

Other thinkers have touched on this fact in their own ways: Philosopher Giorgio Agamben would want us to understand this as a matter of biopolitics, the way that politics writ large is founded on the splitting of existence into two spheres—on one side bare life and on the other life's forms and modes.[180] For Agamben, all classical Western politics itself is a form of government, in that it seeks to order the world, rather than *dwelling in it*. Spinoza, for his view, would say that malicious power always separates the people who are subjected to it from what they can do, forbidding them from effectuating their own powers.

In any event, throughout the secular present and even now into the back loop, this wall first instantiated in Christian pastoral power is maintained in diverse ways, yet its impact on our hearts and minds remains just as powerful: What is good or true can only ever be that which is outside of us, appearing to us as a law or form to be applied, or as the awaiting of an event in the future. We live in glitch mode, our beginning and end never quite in sync. We're taught to live a painful expedition in search of an impossible end, every here-and-now doomed to incompletion, unfulfillment, and obedience. What better way to keep populations under control, lest they rise up against their betters, than by making them sick with self-hate and doubt, resentment and fear?

"Sadness," Gilles Deleuze wrote, "is linked to priests, to tyrants, to judges, and these are perpetually the people who separate their subjects from what they are capable of, who forbid any enacting of capacities."[181] In *Blade Runner 2049*'s final scene [huge spoiler], as snow or soot or ash falls around him, K lays down to die.

"It's okay to dream a little isn't it?" Joi mused just before they passed over LA's massive sea wall, but K knew the truth, "Not if you're us."

In a recent *Women's Health* blurb, a writer clarified that self-doubt is not actually a lack of trust in ourselves. Rather it is a total trust in both our own worthlessness and in an outside power, an undesirable or unpreferred reality which acts as an impediment to us getting a grasp on our own thinking and living. *Women's Health* might be on to something.

Earthbound

The thing about the back loop is that if we take the statement that we are leaving the old world of the front loop behind seriously, this means more than its environmental conditions. Populations and climates are being upended, but we are also leaving the safe operating space of transcendent, sovereign grounds, which throughout the front loop provided the yardstick and measure for what could be done, said, and lived, as well as who could answer these questions and how. God, Morality, Man, Reason, Progress, Law, Truth understood as eternal Facts, Humans as a rational or economic or needy being, even Systems—the power of these sovereign authorities and stable grounds posited in the front loop is coming undone. Every day we tip over new thresholds scrambling what a fact is, what truth is, and what life can be. Passing away as Schürmann already wrote in the 1980s, is not just one standard for thought and action, but the model of referring life to sovereign rules and standards itself.[182] Situated at the end of metaphysics and liberalism alike, "what must I do?" and "what is to be done?" no longer have preset answers. Not only this: they are no longer adequate questions. No more use for deriving "oughts" from political parties, governments, transcendent laws or truths. Referring life to an outside mystery god or rule now appears as just one way of thinking. We are free to begin from our own real. In the back loop, there are no authoritative blueprints.

This intuition has seemingly been taken up recently in the Anthropocene context by French sociologist Bruno Latour.[183] According to Latour, the Anthropocene brings everything once in the background—nature, Earth systems, the nonhuman—to the fore, "smashing" and "dispersing" the key structures of modern thought and life. Included in this destruction is the Human, the modern subject described by Latour as floating in no place and no time like astronauts above Earth. In the Anthropocene, Latour says, there is no more moving from past to future, here to there, to a transcendent place that would absolve or destroy us, provide guidance or foundation, from our given conditions to a "better" or somehow improved world. Instead of modern Humans, the Anthropocene writes Latour, has rendered us Earthbound. Unlike Humans, the Earthbound "behave in a worldly, earthly, incarnated fashion,"[184] "operate under many flags," situated in different territories with their own matters of concern, friends, foes, and gods.[185] "They have abandoned the dream of living under the shadow of any super entity. Secular. Fully secular," writes Latour.[186]

This would seem to be a key insight for living in the back loop in ways other than resilient management. Like resilience, the concept of the Earthbound recognizes something essential about the Anthropocene: old modes of thinking and acting are coming undone. There is a lot to be admired about this idea, specifically its effort to think a form of living that begins from the here and now, a kind of "thought without bannisters" within our shifting epoch, in a pluriverse," as several thinkers have recently put it, of situated, heterogenous worlds.[187] Rather than managing the old operating space, an acceptance that Western civilization is already dead or ending.[188] And indeed the recognition that living in the back loop entails no longer seeking our end or beginning, reason or rule, in a transcendent outside, but rather in the worlds we inhabit. Latour's Earthbound is moreover an important concept not least in that it gives a name to all those forms of life that do not find the real lacking, a waiting room on the way to the true, but as that which grounds life. Latour furthermore rightly notes that earthboundedness is the condition for leaving the abstract arena of classical politics and renewing the possibility of the political as a Schmittean matter of recognizing friend/foe.[189] From this perspective, there are no more universal definitions of the human, but instead situated ones with their own forms of meaning and local context. If transcendent handrails no longer offer guidebooks for life, we just need new imaginaries.

A new sentimental education

What an interesting time to be alive, one might very reasonably think.[190] Considering the effect of front loop structures to dim down possibilities for human life, this would certainly seem like a moment of potential transformation (such after all is the basic premise of back loop ecological theory). Yet what is remarkable in Latour—and as we will see, the broader discourse in which his thought is situated—is how, rather than opening up possibilities for life beyond thresholds, this line of thinking operates to close them down. Rather than offering potential, Latour forwards a set of conditions that are said to define earthbound life in a mandatory and universal manner, and are portrayed as synonymous with a set of imperatives or rules for what is now considered right or wrong in living beyond thresholds. In these injunctions, the previously discussed distrust in and hate for the self—dreams, imagination, and experience—as well ways of separating people from their powers, do not at all disappear along with modern grounds whose end has been so proclaimed, but rather reappear and are extended in new fearful, terror-ridden Anthropocene ways.

Why are we Earthbound? Because we "must," we "have to," it is an "injunction" and "obligation," proclaims Latour.[191] In Latour's view what has "returned" us to the Earth is not *us*—not our choice, a tie to the Earth, or a love for our home—but Gaia "irrupting, hurtling catastrophically at us.[192] For Latour, as for many other Anthropocene theorists, trying to "leave" Earth's limits, to free ourselves from its "shackles of necessity" is what got "us" here in the first place. Critical thinkers almost unanimously portray the structures—and promises—of modernity with scorn or as ruins themselves (to think otherwise, the current discourse suggests, would be out-of-touch with the times—and perhaps worse, eliding or erasing the true nature of the world). Following a similar structure as resilience, likewise do "dwelling in ruins" theorists depart from a key lesson said to apply to each and all of us: "all that taming and mastering has made such a mess that it is unclear whether life on Earth can continue."[193] "Progress stories have blinded us," writes Anna Tsing.[194]

But the Anthropocene, for Latour, puts an end to this "subjective, autistic, anthropocentric" behavior. Proclaiming the epoch of the human, the Anthropocene is said to herald the "end" of the human. Among such discourses and from diverse angles, the Anthropocene is explicitly forwarded as synonymous with the end of human agency: according to one version, the "end of man" has happened *to* us: now entangled in Earth systems, "it can no longer be maintained that humans make their own history."[195] Others, more normative, tell us that Human of the Anthropocene was wrong, that we must send human agency quietly into the night. "Because, no," Belgian philosopher Isabelle Stengers concurs in an interview with Latour, Tsing, and Nils Bubandt, "we are not "a geological force"... We are a power of disturbance, not a force of anything." Latour: "No more agents of history. We all agree on that."[196]

This lesson, Latour tells readers repeatedly, must be learned. Where Humans, Latour says, acted based on hope—projecting from the present into the future, to create—the impetus now comes from perceived threats posed by looming apocalypse.[197] The new nomos of the Earth, under the new "sovereign," for Latour Gaia is today comparable to—if not ("yet") technically defined by—"the legal quality of the *res publica*, of the State, of the great artificial Leviathan of Hobbes' invention."[198] Indeed for him "returning to Earth" is not a matter of going "back to the land" but "the violent *re-appropriation* of all Humans titles *by the land itself*. As if "territory" and "terror" shared a similar root."[199] At the same moment humans are said to be liberated from sovereign Firsts, they are immediately reenslaved to a new force. "Gaia commands, orders, binds as a secular not as a religious power."[200] This power must be obeyed, its imperatives learned. "We should abandon hope," he argues, and direct our art, imagination, and culture toward forcing a taste and awareness for this immense, "hyper" threat and the fright it should produce.[201] He quotes a passage from philosopher Hans Jonas:

> Such an attitude must be **cultivated**; we must **educate** our soul to a willingness **to let itself be affected** by the **mere** thought of possible fortunes and calamities of future generations, so that the projections of futurology will not remain mere ford for **idle curiosity** or equally **idle pessimism**. Therefore, bringing ourselves to this emotional **readiness**, developing an attitude open to the stirrings of fear in the face of the **merely conjectural** and **distant** forecasts concerning man's destiny - a new kind of éducation sentimentale - is the second, preliminary duty of the ethic we are seeking [emphasis Latour's].[202]

A new "sentimental education" in fear, he hopes, will help us to see ourselves as earthbound, captive in the frenetic tangle of the complex, dynamic life support system he calls Gaia:

> *Earthbound*—"bound" as if bound by a spell, as well as "bound" in the sense of heading somewhere, thereby designating the joint attempt to reach the Earth while being unable to escape from it, a moving testimony to the frenetic immobility of those who live on Gaia.[203]

The Anthropocene is said to necessitate a critical framework rooted in our entanglement with complex systems, antihumanist modes of life, and diminished possibilities. Instead of seeking to rise above the Earth presumably to tame it or creating transcendent plans that float above it—for Latour, we who used to believe we could fly must understand that we are Earthbound. Likened by Latour to a "Great Enclosure," in the Anthropocene, "the great impossibility is not being imprisoned on Earth."[204] In contrast to modern Humans, Earthbound don't float in space like dots on a map but rather inhabit entangled "territories,"[205] which he describes as the entangled network of entities—an Uexkullian *Umwelt*—upon which the Earthbound depend for survival.[206] Inverting modernist claims to mastery, Latour argues that "whereas Humans are defined as those who take the Earth," asserts Latour, "the Earthbound are *taken by it*."[207] Here it is not only a matter of showing how human and nonhuman lives are bound up—as have many geographers for example for decades—rather there is a feeling of anxiety and moral panic around the human, an almost hysterical obsession with showing the error of this image.[208]

Instead of *plus ultra*—the once proud motto of "Humans of the modernist breed [who] ignored the questions [of limits] by defining themselves as those who were always escaping from the bonds of the past, always attempting to pass beyond the impassable columns of Hercules"—[209] Latour's maxim for the Earthbound is *plus intra*: giving up hubris and being bound within limits.[210] For him as for the Stockholm Resilience Centre with its safe operating space, terrestriality is a matter of limits: defining them, managing them, and enduring their repercussions. The perfect image of such earthboundedness, Latour says, is the brutal ending to Béla Tarr's film *The Turin Horse* (see Figure 3.1):

> Probably the best (and also the most depressing!) definition of what it is to have shifted from humanity to Earthboundedness. In the final tempest of the last days of Earth, father and daughter decide to flee their miserable shack isolated in the middle of a desperately parched landscape. With a sigh of relief, the spectator sees them finally going away, expecting that they have at least a chance of escaping their diet of one potato a day. But then, through a reversal that is the most damning sign of our time, a reversal that I don't think any other film has dared show, instead of moving *forward* to another land, one of opportunity, full of great expectations, full of hopes (remember *America America*), we see with horror that they come back, exhausted, despondent, bound to their shack, resuming their old even more miserable life until eventually darkness envelops them in its shroud. Those two are Earthbound. They have ceased to be Humans any longer.[211]

Ruins theory

Latour's perspective is unique but not alone. Likewise do several critical theorists argue similarly that, instead of going back, as some hope, and instead of managing the old safe operating space in perpetuity, and finally instead of relying on transcendent plans or rules for what to do, we must shift perspective to see that catastrophe has already occurred and thus rather than awaiting an event coming in the future, we are already "post-apocalyptic." Anna Tsing's work in particular offers a powerful emphasis on imagination to open up new

Figure 3.1: Ending scene from Béla Tarr's *The Turin Horse*, used by Bruno Latour to illustrate being earthbound. Copyright Másképp Alapítvány/Cirko Film/The Cinema Guild.

paths. Indeed as she states, as the old structures "lose traction," it becomes possible to look differently."[212] Instead of railing against our situation perhaps we can explore "look around to notice [and explore] this strange new world," "stretch our imaginations to grasp its contours."[213] This is a valuable perspective in a society often bent on preserving past structures in perpetuity, as is the case with resilience. Here again, however, while emphasizing imagination, across these narratives one finds remarkably *similar* portrays of post-apocalyptic life, with the Anthropocene increasingly seen as reducing human and nonhuman aspirations to survival and safety (the latter understood as guaranteed by resilience and government alone).

For many such Anthropocene critical thinkers, the new "nomos of the Earth" is to "stay with the trouble" and live amidst "ruins."[214] This imaginary—increasingly pervasive in critical Anthropocene theory—instructs readers that they can never be whole. In its fantasies, we are to become like Benjamin's Angel of History, to see modern humanism as "one single catastrophe which keeps piling wreckage upon wreckage and hurls it in front of his feet."[215] With modern humanity portrayed as an error, many such as Eileen Crist argue for foregrounding limits to human mastery and redefining ourselves as "merely-living life."[216] In place of "outdated," "parochial"[217] ways of thinking and being, a new vision of thought and life is forwarded accompanied by universal statements of truth: "we are not in control, even of ourselves."[218] Universalizing and imperative discourses regarding what human being is or should be—in the place of mastery, vulnerability and precarity—as well as universalizing statements regarding "our" responsibility for current catastrophes proliferate. We are now said to find ourselves "living in our messes," as Tsing puts it.[219] In place of what authors imagine to have been "our" former state of stability amidst "summer's easy riches," "global progress's easy summer"[220]— "progress felt great"[221]—for an "us" which could say "precarity once seemed the fate of the less fortunate"[222] since "our pockets are lined," "now it seems that all our lives are precarious... trouble without end... life without the promise of stability... collaborative survival in precarious times."[223] "Broken world thinking"[224] is here seen as the dominant task to be learned, as thinkers such as Lauren Berlant suggest "muddling" or staying with the "hard," "messy," "messed up,"[225] lessons for which can be learned from other peoples who live with ruins or equally from weeds or mushrooms.

This "ruins politics," as we might call it, is defined explicitly as a matter of survival, characterized perhaps best by Nicholas Beuret and Gareth Brown as "salvage politics," a life of survival amidst ongoing social and infrastructural breakdown on a "broken Earth."[226] In these visions, "the Great Enclosure," as Latour concludes, "has been locked up for good."[227] "There is no happy ending."[228] The only thing left to do, we are told, is to "return to the world" in order to inhabit its remaining "cramped spaces"[229] learn to see ourselves as "survivors,"[230] "live in capitalist ruins,"[231] or in Berlant's words, "the wreck of the old good life fantasy."[232] We are to accept a reduced human agency amidst a complex adaptive system of systems, "while agential powers of creativity are projected to the world, the human is reduced to, at best, following the instructions given by the world," as Kevin Grove and David Chandler put it.[233] To have believed anything else was possible, is now said to only have been a dream.

It's beautiful beautiful beautiful

""Am I dreaming?" "Dream?" said Borne." Hailed as a voice of the Anthropocene and earthboundedness, Jeff VanderMeer describes his cli-fi novel *Borne* as a story about "life in the broken places."[234] In contrast to Bladerunner's policing and control, the novel is set in a post-apocalyptic city once governed by the Company, a biotech firm whose live product testing went haywire. Producing chaos and finally the Company's abandonment of the city, this novel portrays a world of crumbled transcendents and government alike. Explains narrator Rachel:

> They had made us dependent on them. They had experimented on us. They had taken away our ability to govern

ourselves. They had sent out to keep order a horrific judge grown ever more unmanageable and psychotic…And in the end, the remnants of the Company had walled themselves off from us when they were done with us, when it became too dangerous…We were on our own. We had always been on our own. We had no recourse, and I cannot tell you how much some part of me had wished to not be on my own, had hoped there would be some person, someone, down in the depths of the Company who would have an answer, who still existed to explain it all, and who, if we asked them to, pleaded with them, would pull a level or push a button to fix our situation, reset it, and bring forth everything afresh. But there wasn't.[235]

Main characters Wick—drug dealer of salvage tech to replace users' painful memories with happier ones lived by others before the end—and Rachel—a survivor, good with weapons, sniffing out traps, growing food in the bathtub—live in an abandoned apartment complex-cliff-face-turned bunker. Secured by sensors and trip wires to protect against attackers, getting by amidst ruins and toxicity while "a ticking clock kept track of the time we had left" as Rachel put it.

Everything changes when one day searching for salvage tech Rachel finds a mysterious object, "dark purple and about the size of my fist, clinging to Mord's fur like a half-closed stranded sea anemone." She brings it home, names it Borne, and as the creature grows the two develop an unlikely relationship. The charred, poisonous worn-out world Rachel called her life—always on defense, knife in boot, trudging, floods, camps, tired people fighting for scraps of old world trash, mutilated bodies floating on water and hanging from street lamps—is transformed as Rachel sees existence through Borne's fresh eyes. Borne's first reaction to the toxic glowing river in the distance beneath their refuge's balcony: "It's beautiful beautiful beautiful." She watches Borne shimmer from fifty eyes to two, from a bottle to a plant, and opens up to him about life before the seas rose, when she had what her people had had what they wanted and knew who they were, how she had wanted to be a writer not a scavenger killer, favorite scenes from novels and how she loved the sea. Hallucinogenic, delirious, Rachel recounts: "I had the sense of things crawling around inside my veins. I was in my bed. I was on the floor." "Shhhh, Rachel. It's just me, Borne." At night Rachel lies awake reliving conversations with Borne that reopened her imagination. Realizing that Wick and she had always been serious and worried, but Borne is filled with wonder and the pleasure of life's tastes.

Together Rachel and Borne explore the world anew. Gazing at the starry night sky: "Diabolical. Deadly. Delirious. Deep." Four new words he'd been trying out." While Wick's away Rachel and Borne explore the apartment complex and the city landscape—just yesterday so rote, but now so alive—racing through the halls not out of fear; "why is this fuuuuuuuuuuun?" exclaims Borne.

Transfiguring the formerly dim-lit, dust-covered world and letting it become magical, Rachel discovers life could be other than survival and bare life. Sensations of her childhood world, flowers, saltwater, hunting for seashells along the beach, feet in sand, parents alongside replace those of rotting flesh, and in her heart of hearts Rachel would give it all up for Borne. "I realized right then in that moment that I'd begun to love him." "Because he didn't see the world like I saw the world. He didn't see the traps. Because he made me rethink even simple words like disgusting or beautiful. That was the moment I knew I'd decided to trade my safety for something else." And yet this "dreamworld" offers a security Rachel never before achieved despite all the trip wires and alert senses. Finding her in pain and afraid, huddling in the crevice of a fallen pillar, riveted by dreams of being crushed beneath Mord's enormous paws, Borne transforms the empty night sky above Rachel; instead of the "dead, poisoned"-looking moon Borne manifests "the tiny flashes and flickers of a thousand fireflies, like the ones on the ceiling at the Balcony Cliffs. A soft, golden blinking from the ground that wished for me to be calm… 'Shhhh, Rachel. It's me'." Rachel realizes: "Borne makes me happy."

But again, was it real? Sadly as with ill-fated Joe replicant and Joi AI, Borne teaches us similar lessons for the earthbound. The reader soon learns that not only is Borne not a person, but worse he is a killing machine produced by The Company (overhanded ref to Google?), designed to taste, sample, and possibly aggregate and report all the "data" that populate the city. "I was made to absorb. I was made to kill. I know that now."[236] Rachel finds Borne's diary and learns the real story. It was all just a dream, Borne just an "illusionist" wielding "tricks", programmed to taste and kill. The happy time they shared—adventures reopening for a weary survivor becoming child again, delight in the world's tastes and colors, the possibility of two beings creating another plane of existence irreducible to this world and its demands—was just "a bubble, a space-time, that just couldn't, wouldn't, last." In the novel's conclusion Wick and Rachel banish Borne, now revealed for the illusionist he is, and the weary pair return to "real" life, getting by among ruins atop Balcony Cliffs. Where K and Joi face another katechon, for earthbound Rachel and Borne, freed of law and rulers—the Sahlins binary collapsed—there is no magical Eden revealed, only a Hobbesian surviving and trudging. Not only are we taught, as with K and Joi, that it was all a dream, with K just a tool of the LAPD or revolutionary replicants. Here there are no such authorities, yet again experience is not to be trusted. Once more, it was all "just" a dream.

The Anthropocene moral code

Borne's deception, K trudging through humanity's ruins, Latour's potato misery. Despite their emphasis on the need for new imaginaries at the end of civilization, in forwarding images of worlds each said to be broken, hopeless, and defined by inescapable interlinked systems, critical imaginaries of post-threshold, post-transcendent Anthropocene times produce a surprisingly homogeneous picture of life and its possibilities. What to make of this homogeneity? Rather than opening up possibilities for life, could it be that by limiting back loop life to these images, despite its own intentions this discourse is instead closing possibilities down? One way to approach this question is to note at a basic level that connectivity, relationality, and entanglement are equally the definitions of life forwarded by cybernetics, resilience, and systems thinking—and the models in which life has been materially reshaped over recent decades (thus dismantling the liberal subject in practice for some time). Might it be that instead of challenging the central beliefs of liberal regimes and the contemporary status quo, Anthropocene imaginaries in fact provide new ways of forwarding what has become its central ideological and technological core?

Indeed there are many similarities with resilience, discussed itself in the previous chapter as the transfiguration of biopolitical governance amidst back loop upheavals. While posed as a critically-minded new political paradigm for the Anthropocene, proponents of earth-boundedness often simply make explicit what is only implicit in resilience, forwarding degraded images of life as insecure, apolitical, and hostage to volatile Earth systems. In these Anthropocene imaginaries implicitly or explicitly beyond-threshold-life is conceived as little different from resilient life, albeit reduced further as hubris is no longer allowed, ideas of future improvement said to be impossible, and creation and audacity denigrated as outdated artifacts of the 20th century. Faced with the shattering of old models of dreaming and creation, authors like Latour us that no more dreams are possible, other than of managing disasters; that no other worlds are possible, other than this ruinous one in which we are enmeshed beyond our control. From critical academic theory to urban governmental planning, the lesson is that abandoning a single—historical, finite, finished—figure of human being also means that we are meant to abandon human agency, projection, creation, and imagination altogether, with the Anthropocene said to necessitate the reduction of human and nonhuman aspirations to survival and safety (with the latter seen as guaranteed by resilience and government alone). Other possibilities—such as those coming from one's life experience or dreams—are pathologized as incorrect, arrogant, or outdated front loop relics.[237]

Taken together with resilience itself, these imaginaries form an emergent apparatus we might name the Anthropocene moral code. Like resilience, this apparatus constitutes a way of *governing* the possibilities now opened in the back loop. This is particularly paradoxical in that perspectives are coming not from governments, but critical and creative voices. Included in any moral code are strict definitions of right and wrong conduct, as well as precepts that are mandatory in nature and universal in their span. Like previous sovereign firsts, this Anthropocene moral code too contains lessons and so too does it teach subjects to distrust their own dreams, imagination, and experience. In some instances, and in particular with Latour, these ways are forwarded explicitly as a lesson, while in other cases via images (such as those of Rachel trudging, etc.).

We are thus brought full circle back to the matter of sovereign grounds to govern life discussed at the beginning of this chapter. It has been the method of liberal regimes throughout the front loop to decide which ways of living are appropriate and which are not and, regarding the former, to mold individuals and populations in this image as much as, when faced with the latter, to eliminate them by whatever means necessary. As noted already, the imposition of a single definition of life—and concurrent denial of the possibility that being is not a fact but a *question*—has long been the purview of liberal regimes but it is not limited to them. From Christian morality to Western metaphysics writ large, the naming of proper being or a single true world located somewhere above, beyond, or below life itself has equally been seen as the task of self-imagined gatekeepers of the real for some time. Yet just as the search for Christian morality over time appeared as an untruth, so now do diverse commentators see liberalism's will to truth as a fable. Such is nihilism as Friedrich Nietzsche defined it: when the long line of transcendent values floating above or beyond the world are devalued, yet one still cannot believe in the world or one's *own* experience.[238]

Rather than getting rid of determinations of being altogether, thinkers in the Anthropocene moral code still define values albeit by reversal. Resilience recognizes the exhaustion of liberal humanism qua

Truth—an exhaustion announced by the Anthropocene back loop, the epoch of liberal Man declared at the moment of Man's dethroning as catastrophe and failure—and responds to this with a new definition of liberal life: insecure and complex adaptive systems and human-nature entanglement. Imaginaries of post-apocalyptic life, on the other hand, despite their important recognition of the fact that thresholds have been crossed and the need for new imaginaries now that old transcendent structures no longer offer models—portray possibilities for back loop life in surprisingly uniform ways as well, forwarding a conglomeration of worlds each said to be entangled, nondualist, non-individual, and characterized by survival amidst ruins. In both cases the *opposite* or *reversal* of what are seen as modernity's erroneous ways are posited: interlinkage with Earth systems not only as a possible way of understanding one's relationship to the world, but moreover as *the* relationship to the world; brokenness and survival in opposition to hubristic modern humanism, etc. In Latour's case, the desire for a new sovereign is as we have seen explicit, manifest in the imagined form of Gaia. From resilience to Anthropocene imaginaries, in both cases the result is that a similar story is actually told: there is only one world. While resilience explicitly seeks to uphold a single world, unfortunately for Anthropocene critical thinkers a single image of life is again forwarded, albeit by thinkers seemingly committed to the opposite of such unitary thinking. In this way, such imaginaries represent not an overcoming, but the completion or perfection of nihilism.[239]

By still seeking to determine the meaning of being—thus ultimately setting up yet another sovereign ground—the moral code works to govern the *question* of being, that there simply is no definition of what being is. We do not know in advance, outside the concrete practices through which people create their worlds. In thinking they know what being is and that they can determine it for others, each seemingly opposed discourse thus washes over the singular irreducible quality of life.

What to make of this? While thinkers such as Tsing clearly seek openings not closures at the end of the world, could it be that for others like Latour, without transcendents nothing else is imaginable other than maintenance and survival? Are such thinkers projecting their experience and perspective onto ordinary people, for whom it never occurred to them to have an "ego" in the first place? Or maybe it is just that imagining any other way of living is difficult while sitting in imperial centers waxing philosophically about the drudgery of menial tasks like collecting water? Who knows, but certainly we are watching the assembling of a new nomos of the Earth. "Sit down, be humble…"[240]

Dwelling in back loop ruins

The disciplinary measures of this new nomos are not limited to the realms of critical theory and fiction. Faced with a situation rapidly outstripping existing models of understanding and action, many seek to reassert firm ground. Just as Christians must submit and live in fear and obedience to God, distrusting and renouncing oneself, in order to find salvation—to think one could find the latter oneself being arrogant hubris—this story is told to us now from diverse prescriptive agencies include such as critical theorists, design firms, and popular culture but also include the realm of journalism and media. This is particularly apparent in the emerging war on post-truth, seen for example in military thinktank RAND Corporation's recent report, Truth Decay, which explores "increasing disagreement about facts and analytical interpretations of facts and data; blurring of line between opinion and fact; increase in volume and resulting influence of opinion and personal experience over fact; lowered trust in formerly respected sources of factual info" while laying out "a strategy for investigating the causes of Truth Decay and determining what can be done to address its causes and consequences."[241] Or consider from a different type of analyst Kurt Anderson's recent article in *The Atlantic*, "How America Lost Its Mind," deploring post-truth and the invention of new realities by lay Americans:

> Little by little for centuries, then more and more and faster and faster during the past half century, we Americans have given ourselves over to all kinds of magical thinking, anything-goes relativism, and belief in fanciful explanation—small and large fantasies that console or thrill or terrify us. And most of us haven't realized how far-reaching our strange new normal has become… Much more than the other billion or so people in the developed world, we Americans believe—really believe—in the supernatural and the miraculous…Our drift toward credulity, toward doing our own thing, toward denying facts and having an altogether uncertain grip on reality, has overwhelmed our other exceptional national traits and turned us into a less developed country.[242]

Anderson calls for the rescue of America from this "fantasy-industrial complex" of conspiracy theorists, UFO believers, ghost hunters, meditators and Foucaultians. To "slow the flood, repair the levees, and maybe stop things from getting any worse," he calls for "a struggle to make America reality-based again" by reinstating a modern model of expert-driven, rational and unitary truth such as that offered by what we all know to be the bedrock of truth, the national news media. Likewise, one need only consider cable news reporters' Trump-like admonitions of anyone who dared set foot outside prior to, during, or after Hurricane Irma: "Serious risks!" "Not smart!" "Flying debris!" "SUV rolls into Brickell, and the passenger has a helmet on! Thinks they are some kind of storm chaser!" The content of dreams can be dangerous and in need of disciplining. Best to chuckle and flatter themselves with the notion that ordinary people's undesirable ways of living or dreaming are nonsensical.

"Sadness," Deleuze wrote, "is linked to priests, to tyrants, to judges, and these are perpetually the people who separate their subjects from what they are capable of, who forbid any enacting of capacities."[243]

The game, then, will end like this: the Anthropocene back loop represents the enlightened recognition of humanity as a sickness, a hubristic cancer on the Earth, a failed experiment that would be better sent into the night. Christian self-hatred transmuted to a species scale for the age of climate change reaches a delirious, nihilistic crescendo. We will watch with perverse pleasure—not dissimilar from the perverse pleasure left and right take in each other's failures—as humans fade into the background, machines and plants now deemed the rightful inheritors of the Earth. Gardens of erotic statues and huge broken human faces will litter Earth's not-so-distant future crust, future fossils that, like the Onkalo radioactive waste repository, will offer evidence of a human species driven to greedy excess, an Ozymandian cautionary tale for the unexpected survivor to find their remains.

The water rises around us. We remain mired in debt and fear. Many succumb to pressure or despair, drugs or suicide. Outside the green zones of Wallace Corporation headquarters, in the dregs of the Company's labs, the masses slog through the churning wreckage of the 20th century, hocking its jetsam in black market stalls, nodding off against the wall of cracked-out projects, soliciting sex and coke, with the only conceivable heroism that of renouncing our dreams and dying alone at the footsteps of a company lab in the falling snow or soot or ash. At best we can toast one another to a Sinatra song like K and Deckard:

> *We're drinking my friend*
> *To the end of a brief episode*
> *So make it one for my baby*
> *And one more for the road.*

Perhaps in time, as Rachel concludes *Borne*, the animal descendants of the Company's mutant creations "will outstrip all of us in time, and the story of the city will soon be their story, not ours."[244] Then, she muses, the city will at last be "truly beautiful."[245]

The power of mental imagery at the end of the world

The thing about stories we tell ourselves and the images we imagine them with is that they are a choice. They are also not without real world consequences. Are we truly just survivors? Is life beyond thresholds necessarily catastrophic? In the fitness world imagination, visualization, and mental rehearsal of both are seen as powerful tools for transforming performance. To do so one not only thinks about something, but creates a rich mental image—detailed down to smell, taste, color, feeling—of a scenario they would like to achieve (running a marathon, completing a deadlift properly, even success in work or family, improving self-confidence or controlling anxiety). As put by Ironman triathlete Ralph Teller, "the more control we have over our imagination, the more we are able to control our performance."[246] "Mental imagery," he continues, "is intended to train our minds and create the neural patterns in our brain to teach our muscles to do exactly what we want them to do." Reversing this logic, can it be that in apocalyptic imaginaries described above we are being subjected to a negative visualization exercise, rehearsing ruins and our own degradation without cessation, thus conditioning our minds to believe such images actually define our realities? Experts on visualization and self-mastery explain that one of our greatest weaknesses comes from not having control over our mental images and imagination, letting it run wild with fear.

My interest is not in critiquing ruins imaginaries so much as to assert that there is not just one way to respond to the end of old forms, the end of a way of life based on transcendent first principles. Dwelling in ruins theories recognize something key to living in the back loop: that the old world is over. We are already "beyond" or "post" apocalyptic, if by apocalypse we understand the end of a certain qualified liberal order of life. But if we are truly living in the ruins of liberal civilization, this need not mean the end. The beginning and the end are one, after all. Why calcify thinking precisely at the moment when it is being opened up? Why reduce life's infinite colors to predetermined frameworks and drab images of survival, as if one knows what living is in advance?

The virtue of the back loop responses I have outlined thus far, both resilience and ruins thinking, is that they recognize the passing away of the old structures and grounds for life. The problem is that they pose themselves as if they were the one, single meaning for such a condition. But there is no single meaning, no one correct response to the back loop. Whereas in the past, politics in one way or another entailed a set of rules or prescriptions for how to live or what to do—a "proper" use—in large part, this too is part of the ruins. There is no answer given in advance. Writes Schürmann,

> It used to be the awesome task of philosophers to secure an organizing first principle to which theoreticians of ethics, politics, law, and so forth could look so as rationally to anchor their own discourse. These points of ultimate moorage provide legitimacy to the *principia*, the propositions held to be self-evident in the order of intelligibility. They also provide legitimacy to the *princeps*, the ruler or the institution retaining ultimate power in the order of authority.[247]

But the time of this way of thinking is ending. Many seek to either freeze the back loop's ongoing transformations or return life to stable grounds, even the new stable grounds of precarity, flux, and entanglement, to which the Anthropocene is said to deliver us. But contrary to continual statements regarding the "obligations" posed to us by the Anthropocene—antihumanism, entanglement, despair—an Earth epoch does not "tell" us anything. If anything, the Anthropocene and its back loop are only the shorthand for a situation, an Earth, a time of intense transformation. As such they can only be taken up, explored, or responded to in different ways. There is no one way to inhabit the Earth or to move on it. There are only the infinite modes through which we enact ourselves and worlds.

If one takes the idea that old world is ending seriously, this opens up much broader horizons. The end of the one world world and its fictive kinships is not the time for reasserting ever new definitions for what life should be, but for reaching out into the infinite range of what we and others might make it. What is needed is not negation of the front loop per se—a countermovement to or reversal of modern civilization that remains caught in what it opposes—but simply the creation of other ways of being in the world. Defining truth and how to live in this context is open to all of the world's masses. It shouldn't be surprising that when ordinary people take these up on their own terms, in their own conditions, what emerges looks far stranger than an AUTOCAD rendering or academic speculation. Rather than being resilient to or in it, rather than defining oneself in these images, even in the negative ("not resilient"), but equally rather than declaring life irreparable, it's possible to simply find new ways of inhabiting the conditions where we are. This entails having the courage and confidence to follow one's own ideas of truth and life, as well as the techniques through which to make them breathe. Such requires thinking life free from both liberal modernity as well as any other imposed models for appropriate being.

Each of the images presented in this chapter are just that—images. They are works of fiction and theory. In the real world, things look very different. Many people do not live this way. Developing and honing the craft of imagining life beyond the limited visions around us --in poetic as well as strategic ways—is part of building ourselves as people capable of bringing into being other worlds. We should trust and forward our own images. The ones we see in our dreams and imaginations, the ones we invent.

Love is also a ground

Anyway as the Stoics said, it's really a matter of perspective. The ground underneath our feet is shifting, the future once imagined, now unknown, and we have a choice: see this as a painful loss and look back with nostalgia, suffering a deficient present to be managed senselessly and in perpetuity. Or, see it as the occasion for things to get interesting, to begin again. Grasping for the safe operating space of the past or a new sovereign might be more comfortable, but there are other alternatives. As Edward Snowden recently tweeted from exile: *don't stay safe, stay free.*

What this means is not a foregone deal. As ecologists say, the back loop is a time of experimentation. For many, just because some forms of dreaming and creating are passing away—thankfully, for many—doesn't mean no other dreams are possible. That so many assume this only reveals the limits of their imagination. Rather what is opened to us now is the possibility of devising our own ways of dreaming and creating.

The test we face is can we stay here, on the brink? Can we inhabit this rift, shape it so that new lands may form? For many the back loop is the end, and no other way of life is imaginable. *Blade Runner 2049, Borne,* and Latour certainly all deliver this lesson in their own ways, but it is repeated in many voices, high and low (the basic training we receive at any Regal Cinema or Barnes & Noble constitutes the new sentimental education by which we are taught the behaviors and rhythms of Anthropocene life: *catastrophe and survival amid interlinked ruins*).

The ancient Greeks accorded great importance to their dreams, seeing them as an oracle that accompanied you across place and time—"a tireless and silent adviser," wrote Synesius.[248] The gods spoke through dreams. But for us living in the undoubtedly messianic time of the back loop, dreams may have a different and more important significance.

We live now in an unsafe operating space, not only because we have already passed so many planetary boundaries, but also because there are no guidebooks, no answers from on high, no guarantees and no assurances. In a back loop the only way forward is to create our own experiments, grounds, and answers—a process that can only begin from the real.

With this in mind, perhaps the real heroes and threshold of the fictional works discussed in this chapter are the love and mutation of Borne and Rachel, Joi and K. What if these characters already have what they need, the answers to the questions they are asking, within the strangeness of their dreams and themselves? What if the most demanding of our attention appears first in what seems nonsensical or absurd—that is, in the realm of dreams, in our interiors and experience? What happens if we each have the courage to turn and face this? Maybe it can release us to become something of our own creation, to dominate the chaotic events that inundate us, instead of being dominated by them.

It seems to me that the future belongs not to those who seek to govern or suffer the back loop, but to those who know what they love, and take that love as a starting point and new definition of security. What we love has nothing to do with a set of external properties, biological or otherwise, rather it is an affirmation of what we live and feel, long for and dream—the physical ground we can stand on, and use to construct dwellings small and vast; new mountains towering into the sky. Or maybe it's a fountain?

> Have you seen the Trevi fountain in Rome? Fountain. Have you ever seen the fountain in Lincoln Center? Fountain. Have you seen fountains out in the wild? Fountain. What's it like when you have an orgasm? Fountain.

Instead of more interlinking—ubiquitously championed as the only respectable reality—could it be that what we need is de-linking? Isn't that what K really dreams of? Choice: Live entangled in ruins, surviving a careening landscape of dust and waste. Or let go, detach from these knots and turn within.

> And dreadfully distinct. Against the dark. A tall white fountain played.

Maybe the back loop is a blossoming idiorrhythmy of diverse and singular realities, and the arts of distance between them.

    Society disdains Joi and K, and movie critics mock their delusions, but to each other they are real. While all the other characters behave with resentment and fear, they follow a love for what they have seen and lived. They begin to believe in their reality and give it shape. K gives Joi an emanator, a device that allows her to become physically mobile, untethered from the apartment where she'd been imprisoned until then. "You can go wherever you want!" She takes pleasure in the rain on her flickering digital skin. In one of the film's most moving scenes, Joi is amazed to see the city and sea wall from the windshield of K's car. Instead of accepting that she's only a program, Joi hires a prostitute so she can sync with her and make love with K. "I want to be real for you," she says to him. K: "You are real for me."

    Finally knowing the risks—she will die if it breaks—she asks K to help her de-link from the apartment mainframe, to become mortal by existing only on the emanator. Maybe K wasn't desperate at all. Maybe he believed in his value and force, and for a moment explored his potential. Joi and K show the possibilities that would be present if we would allow ourselves to trust what we feel to be real. A threesome with a hologram. An adventure beyond the wall and mainframe. Dreaming for anything but a normal day. K asks Deckard if his whiskey-drinking dog is real. "I don't know. Ask him."

Interlude: Getting Out of the Loop

Let's return to the adaptive cycle diagram (See Figure 1.2). Both resilience and post-apocalyptic ruins imaginaries recognize that we have entered the back loop though they interpret our being there in their own ways. For the former, the back loop is a disruptive event that "tests" systems and constitutes an opportunity to try out new management responses. The goal of such techniques, however experimental they may be, is generally to maintain systems' identity or pre-existing state within its safe operating space. In other words, resilience experiments seek to find those thresholds, while testing out new ways to maintain systems. Post-apocalyptic film and theory, on the other hand, sees the catastrophic nature of maintaining such systems, and proclaims that the end has already come. In these imaginaries, the back loop is the world we inhabit—there is no other, they repeat. The underlying reference remains the front loop, with what is left reduced to surviving its remains/ruins—thus always defining life in relation to the front loop past—until "we" humans disappear, and jellyfish or some other entangled meshwork rightfully (in these theorists' view) come to replace us.

Both responses perform a similar blackmail, insisting that there is only one world and one way of living. In resilience's case, existing social conditions of work, debt, profit are backed up by an unending state of emergency, ever-multiplying and expanding network of security techniques, social control, and police measures. By embracing and responding to disturbance events, it is hoped that radical experimentation, such as the raising of miles and miles of roads in Miami, will allow the overarching social and economic systems to remain in thresholds of safe operating space. A mode of governing liberalism's own end, one devised amidst the *realization* that this end is occurring (a realization and fact which the Anthropocene names), resilience channels innovation toward tethering human imagination to the maintenance of existing economic, social, and political relations. Beyond sea walls and wetlands most powerful of all resilience's techniques is its ability to conflate continuation of human life with the continuation of the specifically liberal way of life—to portray the two as necessarily synonymous—and moreover to portray the maintenance of both as the object and goal of each and all.

While posed as a critically-minded new political paradigm for the Anthropocene, the narratives and speculative imaginaries discussed in chapter three often simply make explicit what is only implicit in resilience, forwarding images of life as insecure, entangled with, and hostage to volatile Earth systems. But perhaps more vexing than their proximity with governmental discourses is the way in which such narratives, like resilience, forward a specific, *single* image of life (one moreover which is normative). Beyond safe operating space, all we find are yet more sovereigns and codes for behavior, more theory and art pronouncing nothing else possible but a painful life of survival defined in terms which seek to reverse front loop models. Again there is only one world possible, albeit now *"this"* world, defined as the flotsam and jetsam of the front loop as it careens around us in the back loop. In both cases, for resilience and ruins imaginaries alike, the adaptive cycle becomes a closed loop. There is no getting out. And in this way the *question* that is being—the singular irreducible quality of life, its ability to always give birth to the new—is again forgotten.

But as even resilience founder himself C.S. Holling emphasizes, the first and most important thing to do in a back loop is to, "recognize that we're moving into regimes of the unknown—of the literally unknown."[249] What can happen now, in other words, is not a known or knowable quantity. As resilience thinkers have begun to argue recently, there are many other possible trajectories in a back loop. Refraction. Pulsing backwards and forward. Cross-scale influences and cascades. Forced transformation. Regime shifts. Surprise and novelty, unexpected synergies. Instances where nothing new is created at all.[250] "During such times," Holling observes, "uncertainty is high, control is weakened and confused, and unpredictability is great."[251] And what about the X on the bottom left side of the adaptive cycle diagram, the exit route which appears on all versions of the diagram but is almost never discussed?

Faced with the illegitimacy of current yet increasingly irrelevant societal authorities—political, cultural, intellectual—how to become free? Faced with the untenability of living out liberal life or its ruins in perpetuity, how to develop other ways of living? Instead of repeating it ad nauseum, how to get *out* of the careening, doomed loop?

We've left behind the safe operating space of the Anthropocene's fore loop, thus the back loop is not a brief nightmare event looming at the edge of our vision, but the ongoing now that we are already in. As the great philosopher Peter Kingsley has said: we all sense, intuitively, that liberal civilization is already finished. The thing to do this situation is not, he suggests, to keep things going desperately, running and running with nothing beneath our feet but thin air. The end, Kingsley argues, is the most important part. As anyone who's ever participated in a ceremony or ritual knows,

> Nothing works unless the finish is perfect. It's the final move that finishes everything off. If something is left uncompleted, it's a complete waste. That is actually what is needed and called for, from a real spiritual point of view, at the end of a civilization. Not that people go on the way they've been going on for the last few hundred years but that they help consciously to bring that civilization to a close. Respectfully correctly, completely, with dignity.

Endings need not be lived as tragedies. In this case, with the world ending being that of liberalism's world, one many of the planet's population has been trying to escape its whole life and indeed for generations now, this may be especially true. Entering into the back loop need not only be understood in terms of listicles of destruction wrought by a homogeneous human geological agency but ought to be understood as a name for the twilight of liberalism, its single world order and fictive kinships coming undone and the opening to other possibilities that this unraveling permits. Kingsley concludes:

> It's *beautiful* that it's over, but it needs to be made consciously over, to clear the way for the new, for the future… [Our] job is with dignity, discretely, to bring things to a close. To wrap up what needs to be done here. It's a beautiful task. This is 'pulling up the tent pegs,' taking down the tent. It's a wonderful time! Our business is here, it is now, and it is a mystery.[252]

Letting it end allows us to get on with the beauty and mystery that is living. From this perspective, the back loop is a time to evolve and shed what no longer serves us in this moment. To try things out. See what feels right, what fits. Instead of sentimentally taking refuge in the past, and rather than looking to de facto but out-of-touch authority figures, we now have the opportunity to redefine living and thinking on our own terms.

From this point of view, life and future are radically open, and we find ourselves in an unsafe operating space. This is not another reference to risk—we are already inundated enough with that—but to the fact that leaders, sovereign principles, and political handbooks no longer dictate life and thought. As ecologist Lance Gunderson puts it: "we can't analyze our way out" of a back loop; "the only way is to probe uncertainty."[253] Put otherwise, as we explore our own possibilities here, we find ourselves without grounds: no transcendents, no answers from on high, no guarantees. In this sense we may be terrestrial, but not in the sense described in the previous chapter. We don't have to answer questions in the old ways. In fact what is certainly also on the table is what are the right questions themselves, as well as who can answer them and how. The back loop is a time to explore, to let go—of foundations for thinking and acting—and open ourselves to possibilities offered to us here and now. In such a situation of the extreme unknown, Holling suggests, "the only thing you can possibly do is experiment."[254] Instead of accepting the end of human agency except that of managing crisis—and rather than imagining ourselves as victims or managers of the back loop—in the remaining half of this book I argue that another possibility exists: deciding for oneself, locally and in diverse ways, where and how to inhabit the back loop.

While critical theorists like Latour proclaim "our" unpreparedness for a terrestrial existence, counseling diminished expectations and diminished horizons—thus locking us into a similar blackmail to that forwarded by resilience, one which says there is only one world, one future, and one degraded subjectivity—for many outside of academia's hallowed halls, the Anthropocene and its possibilities look very different. Climates are changing, seasons shifting, and habitable zones for people, plants, and animals moving along with them. As the structures of the Anthropocene's ascendant phase splinter, matter and energies are released, opening to new potentials. Indeed, the Anthropocene's back loop is also marked by wave of experimentation with ways of living in transforming environments—ecological, social, political—including beyond perceived thresholds or safe operating spaces. This experimentation does not seek a return to an imagined "before" the back loop nor its mere continuation, and neither still is it geared toward simply surviving its ruins. Where many fear the back loop, these experimenters are comfortable here on the brink and already shaping it in their own powerful ways. For them, just because old ways of being hubristic and living are passing away does not mean that agency is dead or that no other hubris, and no other living, is possible.

Such people are already inhabiting the back loop as they experience it variously across place and situation, in very different ways than those said to now define life. The practices they are testing out offer ideas of what changing life in the back loop could mean, and as importantly raise some key questions of the back loop. These include but are not limited to: How to live with water? What is beauty? What is health? What other forms of subjectivity are possible? What are other ways of seeing and perceiving the back loop can be invented? What tools, techniques, and methods are needed in each case? Looking at how ordinary people are exploring such questions will also offer ideas regarding what methods and tools may be useful for inhabiting the back loop, as well as what forms of transformation are now on the table.

Each story that follows is just a story, of specific local phenomena. What runs through them however is a different way of inhabiting the back loop, not through discipline or governance, but through the free use of tools and techniques suited to peoples' own needs and desires. Rather than offering imperative statements or laws to which life must constrain itself in order to survive, they instead communicate to us that life in and beyond the back loop is something to be explored on one's own terms, via local, situated practices and free use of tools.

4 Survival Skills and Floating Houses

Disaster prepping and survival skills

In November 2012, just after Hurricane Sandy, a New York City Preppers Network Meetup filled the meeting room of a church on 189th street in upper Manhattan. Perhaps a hundred people of extremely diverse race, age, and gender packed the room, sitting in folding chairs or perching on ledges to spend the afternoon sharing post-storm know-how. One by one attendees stepped up to the folding table at the front of the room and showcased favorite items in their bug out bags: water filters, life straws, and UV pens; tarps for shelter; fire-making tools like flint, waterproof matches, lighters, and artificial kindling; sanitation supplies like soap and toothpaste; first aid kits; 2-way radios, walkie talkies, mesh network devices; compass and both digital and print maps; solar flashlights; basic multipurpose tools including knives and bandanas.

Organizing the meeting was 39-year old African American firefighter Jason Charles, the NYC Preppers network founder with a thick New York accent and wearing paracord bracelets, whose own bag includes 2 weeks of Meals Ready to Eat (MREs). By and for urban dwellers who for the most part live in apartments, rather than homes with garages to store their gear or rural tracts of land for farming, the NYC Preppers Network takes up the challenges of learning and building preparedness for food, water, and sanitation in the urban environment. "We're not talking about alien invasions, we're talking about realistic shit that can happen, and has happened," clarifies Charles.[255]

Over the past decades, natural disasters and hurricanes in particular have posed key questions of the back loop. In New Orleans after Katrina, thousands were trapped in the Superdome because they had no place else to go, and police even prevented some people from leaving the city at gunpoint. Blackwater helicopters serviced wealthy neighborhoods, while poor people died on the sidewalk with no clean water to drink. While Goldman Sachs and New York University had power throughout Hurricane Sandy, working class areas of Queens and Brooklyn were in blackout for weeks and months. After Category 4 Hurricane Maria hit Puerto Rico in 2017, the entire island had no running water, little working communications infrastructure, and a near complete lack of electricity.

The story is similar in each micro back loop: Green zones for the rich and Superdomes for the poor. These disasters have moreover been taken by governments and corporations as opportunities to not only reassert control over existing conditions but also to remodulate and refine them in real time. In addition to the resilience industry, here think bitcoin millionaires building communes in Puerto Rico,[256] government and NGO contracts a la what Naomi Klein calls the shock doctrine, as well as a growing disaster relief industry.[257] Neoliberal and resilience institutions also preach community preparedness. *The New York Times* now owns a new company, The Wirecutter, dedicated to testing and marketing the best emergency preparedness supplies (which they regularly crosslink in articles about disasters) including water, food, first aid kits, flashlights, extra batteries, personal medications. A reverse *Bowling Alone* is encouraged, with individuals seen as less resilient, and tightknit communities—where neighbors know each other and share skills and supplies often networked via local hubs like stores, PTAs, sports clubs, or churches—especially encouraged. Awareness of risk and possible disasters are to promoted: The Department of Homeland Security (DHS) Ready.gov offers information on 28 different kinds of disasters including bioterrorism, cyber incidents, active shooter, drought, explosions, landslides, home fires, wildfires, space weather, and volcanoes—as well as learning about socioeconomic vulnerabilities—alongside Community Emergency Response trainings encouraging communities to organize into teams. And of course from million dollar secret escape routes out of NYC to billionaire bunkers in Kansas, the wealthy are deeply invested in prepping too.[258]

That being said, for preppers and others, disasters have brought up important questions of a different nature: how to not be hostage to relief agencies, FEMA camps, or governments that disdain whole populations? How can one help oneself and each other? How can you have a greater degree of power over your situation? What skills and knowledges does one need to learn now to do so? As well as the desire to become heroes, learn new skills, and decrease dependency.

While answering these questions has led many people to disaster preparedness, countless others are talking about these matters not just in terms of disasters. For a multitude of reasons, growing numbers of ordinary rural and urban Americans of diverse origins are taking up learning skills.[259] Attend survival skills classes in upstate New York or deep in Queens, and you will be sitting on wood logs sharing a fire with Dominican dads trying to help their kids learn about life; white and black suburban families with teenagers in hot pink bandanas; kids from Flushing leading field trips to the woods to help their fellow youth toughen up; all trying to skill up on basic life capacities they see themselves as lacking. From classes to gear, neighbors and

Survival Skills and Floating Houses

Figure 4.1: Jason Charles, founder of the New York City Preppers Network. Photo by Jason Charles.

individuals are organizing to help each other and themselves in the next storm or backing up communications case of network outages via landlines, satellite cells, or even HAM radio training.

While popular media and academic literature often stereotype preppers as a mixed bag of white racist men or conspiracy-obsessed wingnuts, there is a much broader participation of races and genders and a deeper interest than personal self-preservation.[260] Some are certainly religious zealots. Some are patriots, some right wing, "Bible-believing Christian conservatives," others are "alfalfa-munching Birkenstock-wearing leftists" as former US Army intelligence officer and survivalist James Wesley, Rawles puts it, adding, "the more the merrier."[261] Whether motivated by fear or by love, across this they are people who've gotten prepared—for whatever it is. As one Florida-based prepper puts it, whether he stays in or bugs out, in a disaster situation he knows he will "encounter people with whom he doesn't agree, preppers of a different political stripe. Anarchist hippies, communists, gangsters. He doesn't mind, as long as they're prepared." His response to them through decades of prepping is simple: "I guess I'll see you out there."[262]

For many, prepping becomes a whole lifestyle lived not in abeyance but in the here and now. There are weekend retreats to experience life in the woods with just a group's own supplies; bug out simulation walks from Brooklyn to New Jersey (in winter cold and summer heat); group buying clubs to minimize gear costs; workshops from canning and bushcraft to How to ____ (tie a knot; start a fire; dispose of a

Figure 4.2: Learning to make bow-drills, Mountain Scout Wilderness Survival School, Garrison, New York. Photo by author.

corpse in an epidemic; navigate by star...) or How to Make Your Own ____ (cough syrup, soap, toothpaste, Altoid tin first aid kit, etc.). As in the show-and-tell event, much energy goes to debate (UV pen vs. life straw; best way to xx) and the sharing stories of one's own experience ("In Katrina ..." "Yeah in Maria we got water from the river for washing and the toilet"). What you wear, the route you take, whether you bug in or out, and what gear you need all depend on weather, time of day, season and even the event type itself, not to mention immediate surroundings like people, traffic, and so on. There is no one-size-fits-all bug out plan, as each person or group is expected to create their own as appropriate to needs and environment. Some preppers see learning to care for oneself and others as a basic trait of maturity and dignity. Some distrust or fear government (it is an absolute basic assumption that no preppers trust government or media). Others are curious, motivated by the joy of learning. Others are on a quest for something, whether dignity, autonomy, security, or a degree of knowledge about or connection to the conditions of their existence.

While prepping can take on its own "long now," circumscribed disaster events have operated for better or worse as real time trial runs for ordinary people to use their skills, test what they can do and discover just how much they have to learn. When Katrina happened, Louisianans as well as thousands of Americans—who initially saw the devastation on television via Kanye West or Anderson

Cooper—traveled to New Orleans to help with recovery, sharing Tyvek suits and respirators with hundreds of other ordinary people on the ground gutting homes in neighborhoods like the Lower Ninth Ward. Likewise in New York, an outpouring of self-organized disaster relief spontaneously emerged in the wake of Sandy. Across the city, youth on bikes, parents with kids, and hipsters alike showed up in droves to shovel debris and help cook for hundreds in hardest hit areas like Staten Island or Rockaway, while many others opened their businesses or homes to neighbors as makeshift organizing centers. Like the guys who had a grill on back of a pickup truck after Sandy driving around giving out burgers, so many New Yorkers realized they could just directly take care of each other. More recently, during Hurricane Harvey, residents transformed inner tubes, inflatable mattresses, kayaks, paddleboards, motor and airboats into all kinds of floatation devices for rescue and escape, delivery of goods and of each other to safety without waiting for government aid. In a more coordinated fashion and at an impressive scale, the Cajun Navy, a network of Louisianans and now others, have used their boats to rescue stranded residents from Florida to Texas since Katrina in 2005.[263]

Prepping and disaster response are united by their interrogation of infrastructure, the texture and support system of our lives, a form of power so clandestine we often do not see it. Armies and insurgents target local infrastructure to take down an enemy; authorities raid informal connections; and insurgents both target urban infrastructure while equally building their own. In non-war conditions, infrastructure is a form of warfare waged by liberal regimes to shape everyday life—how we eat, get around, light our homes, even our ideas of happiness—and prevent the emergence of other ways of living.[264] When preppers and responders take up questions of survival or recovery amidst the breakdown of infrastructures to which we are accustomed, they are immediately led to questions such as: what are other ways of preparing food for hundreds or thousands? Street turkey fryers and barbecues are good for a few days, but what about the longer term? How to produce, share, and store food? Likewise how else can we stay warm? How to power the devices we need, turn off the ones we don't, and keep some lights on? How to obtain clean drinking water? Such questions get at the heart of any transformation in ways of living. This is particularly important if instead of cycling back through the single, homogenous loop, we wish to get on with other ways of living.

Figure 4.3: Cajun Navy, Hurricane Harvey, Houston, Texas, 2017. Jason Fochtman/©Houston Chronicle. Used with permission.

Okay, why am I working for that guy?

Outside the frame of disasters, similar questions have motivated Open Source Ecology (OSE) to create a "civilization starter kit."[265] Started by Marcin Jakubowski—a Polish-American fusion physicist who after obtaining a PhD realized he couldn't fix his own tractor—OSE has grown from two people living in a crude, mud-brick hut to a larger project involving on-site and online collaborators including programmers, mechanics, engineers, and individuals who for whatever reason want to change the world. OSE has a compound on a 30-acre parcel of land one hour outside Kansas City, Missouri—the Factor E Farm—with on-site living units and connections to projects like urban farming. The ultimate goal of their now-rebranded "Global Village Construction Set" is to assemble the 50 to 100 machines most necessary for anyone anywhere to create their own civilization from scratch, including circuit makers, bread ovens, tractors, 3-D printers, eco-housing modules, automobiles, and an interchangeable "power cube" (see Figure 4.4). Open source and affordable, the goal is to distribute and make the information and technical know-how accessible to anyone, who can then build, mod, or even sell the machines to make money.

Their first machine, the Compressed Earth Brick Press, can be produced in one day, and they aim for the same one-day production time for all fifty, as well as fabrication of automated machines that produce other machines, designed for a few hours per year maintenance.

Survival Skills and Floating Houses

Rather than making single machines, the goal is to create "modular, scalable construction sets for building any machine"—burned onto a DVD. This "civilization reboot experiment" is documented in micro-detail, resulting in an online presence of spiraling, almost incomprehensible detail, including YouTube video diaries, TED Talks, wikis, and blueprints.[266] They offer internships, extreme design/build summers, network and prototype via Design Sprints on Google Hangout, and constantly fundraise, with hundreds of "True Fans" now sending $10-100 a month and other crowdsourcing platforms. On an ever-amassing YouTube library, Jakubowski and colleagues in coveralls document the project's successes (a building's foundation was laid) and failures (two machines broke laying the foundation) from amidst open circuit boards and piles of bricks. Concrete trucks roll in, CNC routers whirl, welder sparks fly.

OSE faces the same problems as all small projects led by obsessive/visionary maniacs: the burden of maintaining vision and organization falling on the shoulders of one or a few people, the need to shape nearly everything said or produced to generate publicity, likes or followers, as much being reliant on crowdsourcing funding, etc.

Nevertheless, in their eyes, at stake in this attempted mass redistribution of the means of production, in Jakubowski's words, is "creat[ing] a cultural revolution—where you can Build Yourself—and build the world around you." In a time when many things are opaque by design—technical knowledge monopolized by experts or corporations—and even sharing of ideas in academia avoided for fear of theft, for Jakubowski this technical capacity is understood in his own words in terms of

> the raw power this gives to people—to tap autonomy, mastery, and purpose—towards rebuilding their communities and solving wicked problems... True freedom—the most essential type of freedom—starts with peoples' individual ability to use natural resources to free themselves from material constraints—to unleash human potential.[267]

Imagine if you had these means, he says: "well that will change a lot of different things. It's like you're gonna say, okay, why am I working for that guy if I can actually in my community produce effectively to meet my needs?"[268] The long-term implications OSE sees in these shifts are even more transformative in scope: "Governments as we know them become obsolete. We foresee an equal playing field of competent, well-organized, small-scale, decentralized republics after the borders of empires dissolve."[269] "Let people have the choice," he says. "If they want this, they can have it. If they want the same old,

Figure 4.4: Open Source Ecology's Global Village Construction Set, an open source plan for 50 industrial machines needed to start a small civilization. Open Source Ecology/Creative Commons Attribution-ShareAlike 4.0 International.

same old as today, they can have that. Personally I think this can be way more compelling."[270]

As for the millions of people by themselves or in groups taking up survival skills and prepping, control over one's own life is a central desire here, a desire some theorists, mentioned previously, suggest we throw away. Though prepping is often tethered—albeit often minimally, almost as an pretext—to awaiting an exceptional moment when everything would change—a storm, apocalypse, whatever—in fact the temporality of OSE is in more appropriate to the back loop. The world as we know it is already ending, and there is no need to cling to structures from the past that no longer serve us. For OSE, leaving current modes of social and economic organization does not mean going back to any primitive past, and it is mature enough to recognize the many incredible feats of industrial society. The vision is thus putting high-tech tools for building and designing new civilizations "with modern comfort" in the hands of anyone, including 3-D printing or sleek eco design. It's one thing to train and design for intermittent disasters, flooding, and infrastructural breakdown. But the back loop poses the need and possibility of transforming the very modes of our lives here and now.

Living with water

Let us take this one step further. Sea level rise and coastal flooding is an obvious such back loop challenge. Many say humans cannot live beyond thresholds—more specifically that water will lead to evacuation. But consider the working-class fishing community of Old River Landing, two hours north of New Orleans. Located outside the Morganza levee system on an old bend of the Mississippi River, the area floods yearly with overflow from the Mississippi inundating the collection of some 200 structures and fishing camps, for weeks or sometimes months, with flood frequency and duration increasing in recent years according to residents. When I visited in April of 2018, Buddy Blalock met us at the Post Office and led us on dramatic a drive atop the levee, to the community's entrance. The sign at the entrance reads, "Thursday Steak Night," and beneath that, "Freedom Isn't Free." As we head to his own house, he tosses us life vests: "we're gonna need to take a boat ride."

In the face of more flooding, many residents have gone amphibious rather than leave. Drawing on the region's tradition of camps and swamp living, but using modern hurricane proofing and industrial materials, they outfitted homes, trailers, and even a full bar-restaurant with support poles and Styrofoam blocks, allowing structures to float off their foundations while preventing lateral movement and diminishing wave action. In non-flood times the neighborhood is filled with ATVs, fishing boats, playing children. Some residents maintain gardens in their front yards, and the summer is marked with a July 4th fireworks display. During flooded times, when the river crests to 30 feet or higher, the residents that remain use small boats to get to their homes and camps from the levee under a Spanish moss-laden Cypress canopy. As Buddy, a retired computer man and one of the first to go amphibious, puts it, "some people think this is a problem. I don't think it's a problem. I'm on a permanent cruise," he tells me with a wink.[271]

While management agencies deem living in the flood zone dangerous—with US Department of Housing and Urban Development even launching a controversial $48 million resiliency experiment to relocate all Biloxi-Chitimacha-Choctaw and United Houma Nation residents from sea level rise-inundated Isle de Jean Charles in a coastal area of southern Louisiana, making the Native American tribes "America's first climate refugees"[272]—for the fishermen at Old River Landing, water is instead a problem to be worked around. They drew up plans based on the weight of their belongings and house, watched YouTube videos, learned from each other's models, and got parts from friends in the oil industry. Says amphibious restaurant owner Jacques

Figure 4.5: Amphibious home of Buddy Blalock, Old River Landing, LA. Photo by author, 2017.

Lacour, who is Cajun, "I mean, it's a sit down with coffee and calculator kind of math. But it's not complicated. You do the measurements. And people have been doing this since the 1800s, although I don't think they had the advantage of Styrofoam. Everyone says, 'I don't know why the professors think this is such a big deal'."[273]

Rather than giving up human agency or a form of resilience conceived as survival among the ruins, Old River Landing is a story of people who love the part of Earth they inhabit, a kind of living many argue is no longer possible. Residents—some full-, others part-time— have taken up the challenge on their own terms, neither following a blueprint or answer from on high. From tracking river levels via smartphone apps to actually designing amphibious architectural plans, solutions were developed in an ad hoc way to local problems, tested and trialed in reality. In that they do not important abstract frameworks, nor wait for salvation in the future, but begin from their own contexts, Buddy, Jacques, and the other inhabitants of Old River Landing are certainly terrestrial. But they are not bound to the Earth, at least not in the sense meant by Latour.

Although unfortunately they will probably be read this way, it is insufficient to describe these practices as exemplars of the resilience of the poor, their ingenious ability to live precariously amidst negative conditions imposed on them and which are deemed inevitable.[274] From this easy reading, such practices constitute the perfect neoliberal

Survival Skills and Floating Houses

strategy for abandoning communities to care for themselves amidst the volatility generated by the very same forces that say, you can never be secure (such readings moreover assume that being administered by neoliberal regimes is in all cases preferable to caring for oneself, an assumption that should be questioned). Like Latour, resilience requires subjects who do not argue with what is, but accept and adapt to it, actively exchange all forms of security for insecurity, and who in their own living confirm the validity of images of life as vulnerable, insecure, and paralyzed to survival amidst catastrophic ruins.

Neither, finally, is this a matter of entanglement nor becoming posited as an ontological or political "good" or "must." For example the fishermen may ask themselves: Are the solunar tables on point this week? Which side of the river are the bass on today? In asking these questions and in living their lives residents are at the same time in a variety of relations with forces other than human—sun, water, fish—but this is not a matter of exemplifying a particular political ontology but instead of the actual practices in which they are engaged and what those practices allow them to do. Such criteria, as well as the happiness or satisfaction which comes from living in one's chosen way, are what make a "good life"—not choosing the correct side in a false dichotomy of being separate from or imbricated in more-than-human entanglements.

Such experiments are in fact better understood simply as practices of common human beings as they freely take up and define their lives through experimentation and free use of their transforming environments. Rather than accepting entanglement in the given order of things as is—flooding=moving=dependence—such experiments are a testament to how diverse people operate; without transcendents and in ways quite different from the models forwarded by experts.

Not only did the residents in Old River Landing make the choice to chart out their own path, defying predictions of human behavior faced with climate change. They also make *free use* of their environment, projecting themselves over and against it in order to live with it. Likewise, these experimenters weave traditions and tools into new arrangements that suit them. Their situation is not chosen, but inherited. Yet rather than accepting these conditions as the limits of their lives—ending the story there, or defining themselves by their problems—residents have made use of other resources in their toolkits in order to step back—*out* of "entanglements"—and through meticulous assessment of their conditions—ongoing and modulated each day, via phone weather apps, local trial and error—carve out a place for themselves in their environment, in and with it. That is, the water transforms their lives, affects and radically alters them. But they also

Figure 4.6: Jacques Lacour, Old River Landing, LA. Photo by author, 2017.

assert a place for themselves within its ebbs and flows. In doing so, they make themselves less vulnerable. (Well! We attached Styrofoam blocks to the base of our homes. Now we're not vulnerable anymore.) As Lacour said, "it was just common sense. An obvious solution to a problem."

The problem in this case is not just that of how to get by amidst negative conditions. Rather, it is how to continue what for the fishermen is their definition of the good life, the lifeway they have chosen, on their own terms.

Infrastructure

In the back loop, things are changing and must change. This is indeed a situation of the fundamentally unknown, a fact attested to in the new colors and charts created to indicate unprecedented rainfall quantities, and new storm names. Writes French theorist Isabelle Stengers, "if the epoch has changed, one can thus begin by affirming that we are as badly prepared as possible to produce the type of response that, we feel, the situation requires of us."[275] No wonder we are dumbfounded, writes Latour. But here once again in contrast to such stories told to us of fear, vulnerability, and terror, in reality from post-disaster scenarios to drastic environmental change, humans have much experience with threshold times.[276] And we have long experimented with fire, water, shelter and food.[277]

As is each historical moment, ours is a new and singular time. As we exit the front loop and its single world world, depending on

Figure 4.7: Buddy Blalock, Old River Landing, LA. Photo by author, 2017.

where we are we find ourselves with new questions. Today inhabitants of coastal cities such as Miami are asking how they will live in a flooded city. What happens when the housing bubble bursts again? How to obtain clean water or deal with sewage if salt water intrusion disrupts existing infrastructure? And maybe one of the biggest questions of all, food. Or more darkly, think of Fukushima, the thousand-year half-life of various radioactive elements now dispersed across the islands of Japan. This very quickly brought up material questions for people living there, not simply of how to shut down the nuclear power stations, but, how can they live with nuclear contamination?[278] In Japan post-Fukushima people now say, "we want to choose the way we die" but also "we want to choose the way we will live." Describing the effects of Fukushima one resident who relocated to Fukuoka, Motonao-gensai Mori, recounts:

> Our way of life collapsed. Invisible toxins travel through the food chain. The government tries to mix radioactivity with cement, force populations to eat locally grown vegetables and fish; trying to build an ice wall around Fukushima Daiichi reactors. But there is a positive aspect to this. People are changing life. Office workers becoming farmers. Teenagers learning to hunt and trap. Neighbors opening markets to trade what they're producing now: animals, crops, haircuts, rice rolls. Some talk about preparing infrastructure in the west so more people can come join. I want to become like Bear Grylls now.[279]

According to counterinsurgency consultant David Kilcullen, the future will be crowded, urban, coastal, and connected.[280] While it is possible that today's political leaders, business, and associated institutions could create a few "Super Venices" supported by sea walls, pumping systems, and ecologically resilient infrastructures like oyster reefs or reflooded wetlands, it's guaranteed that such cities will be dedicated to preserving spaces for the elite while most everyone else bears a kind of diminished existence, increasingly bearing the brunt of rising seas, infrastructural disruption, and so on. All of this, of course, while making sure the majority remain dependent on work. If we ask instead "Is our political system capable of producing a *desirable* or *livable* outcome for humanity?" the answer is absolutely no. That being said, despite all rhetoric to the contrary—that we are helpless, isolated, powerless—regular people have incredible capacity to handle themselves. Ultimately, this is now our opportunity to stake out entirely new possibilities for ourselves and each other.

While framed in terms of disasters and environmental matters for the purposes of this chapter, none of these questions need be oriented around such scenarios. The back loop is not limited to environmental or disaster matters, but concerns the coming unhinged of the very structures of liberal life itself. Deciding on one's own terms where to go from here, can everywhere be a matter of taking infrastructure, architecture, and design in one's own hands and wielding them as the powers they in fact are.

The simple fact is that infrastructure is key for living in and especially beyond the back loop. This is so in multiple ways. On one hand, liberal politics and power are infrastructural, central to the governing of populations and environments.[281] Understanding the constraints on us requires understanding that liberal governance does not reside exclusively or even primarily in governments but consists rather in ad hoc assemblages of technologies and designs, architectures, and infrastructures that make up the built environment. Infrastructure constitutes what Keller Easterling calls "extrastatecraft," "the secret weapon of the most powerful people in the world," the "unstated" "content manager dictating the rules of the game in the urban milieu."[282] From this perspective, infrastructures are powerful political devices because often they do not appear to be doing anything and even appear natural or simply "there." However, alongside their seeming background role, infrastructures also inspire powerful visions and fantasies themselves in diverse relation with government's own political imaginaries.

As discussed in chapter 1, front loop infrastructure was a means through which the historical anomaly we call liberal existence was reentrenched and extended, modified and recalibrated, through the creation and functioning of everyday space, architecture, movement, and human relations. The term "infrastructure" itself originated around the turn of the century as railroad engineering jargon: "the tunnels, bridges, culverts, and 'infrastructure' work generally of the Ax to Bourg-Madame line have been completed."[283] In the United States, as elsewhere, the word developed greater usage in postwar civil defense and urban planning, appearing as military logistics language in NATO's 1950s "common infrastructure programs," in which member countries pooled their money to construct the various military installations—communications, airfields, war centers and training facilities, fuel supply systems, pipelines, radar systems, ports, etc.—necessary for modern, omnipresent warfare.[284] More recently infrastructures have been important in post war reconstruction efforts in a place like Iraq, where the ability to get them up and running was the litmus test not only of U.S. power but also of the form of life it promises. In these scenarios, as Eric Schmidt and Jared Cohen of Google argue, a "communications first, or mobile-first, mentality" has emerged wherein the reestablishment of communications infrastructure has become the first priority in the long process of rebuilding entire societies, providing a "new cement" that is not only a strategic objective but also a method of counterinsurgency.[285] Poetic as well as technical, front loop infrastructures also enabled new experiences and perceptions, transforming imaginations and forwarding powerful desires and dreams.[286] From clothes irons, radios, and refrigerators, to air conditioning, running water, and electricity, front loop infrastructures of liberal living were intimately tied to redefining the meaning of happiness or success, as much as they reshaped how people ate, traveled, and communicated.[287]

Today however infrastructure is increasingly coming to the fore due to the growing frequency of its failures, including disasters like the meltdown of the Fukushima-Daiichi reactor, the Flint Water Crisis, Hurricanes Katrina, Sandy, or Maria, as well as blackouts and terrorist attacks. Infrastructures have been increasingly recast as bulky and brittle systems incapable of surviving a world of complexity and volatility. The once glorious notion of order promised by infrastructure is now derided as an artifact of an exhausted and imploding humanist epoch. And front loop infrastructure, once thought the pinnacle of the era, is now seen as epitomizing a fatally flawed idea of hubristic human mastery and the cause of today's cascading damage, the latter a sign of a future in which crises are projected to worsen in scope and severity as climate change and its effects progress.

Scientists heading the Anthropocene Working Group—the subset of the International Commission on Stratigraphy tasked with determining the validity of naming the new epoch—have centered much of their research on existing infrastructures, which are examined in the present as "future fossils."[288] As discussed in chapter 1, geologists measuring the Anthropocene now date its start around 1945, part and parcel with the introduction of many of the once-celebrated front loop infrastructures.[289] Mines, digital networks, and industrial agriculture alike are identified collectively by scientists as part of an expansive "technosphere" which was brought together in the post-1950 Great Acceleration to join the biosphere or lithosphere.[290] This "machine room of the Anthropocene" is not seen as a feat of Western civilization garnering awe and faith but a careening threat that must be managed.[291] In it, what governments and companies now seek are ways to defend complex infrastructural systems not only from the outside but also from themselves.

In this shifting context infrastructure has seemingly taken on a new meaning. While infrastructure of the past was imbued with ideas of progress or the promise of a (better) future, the vital systems that we are now told include us project an image of existence permeated by crisis, of an ever-expanding universe of tipping points. This cobbled together and tangled web—now stretching to the ends of the Earth—of flows of people, water, and energy, of wetlands, code, and satellites, has only the aim of managing or surviving its crises. In this, we are not only told to be patient and prepare to undergo whatever crises to come, crises that are an acknowledged byproduct of the self-same system. As discussed in chapter 2, as planners, architects, and city officials attempt to make cities resilient to the myriad ever-increasing categories of risk, life itself is being recast as infrastructural in nature, and infrastructures are cast as indistinguishable from life in the Anthropocene.

On the other hand, however, in terms of living freely in or beyond the back loop, infrastructure is being, and can be, redefined. As many existing infrastructures increasingly appear as problems or hindrances, locking populations into a crash course trajectory, the reclamation of infrastructures and skills has become increasingly key for living in the back loop. And while it is useful to recognize the role of infrastructure in governing populations, its exploration is in reality the purview of everyone, and not just resilience planners or militaries. Just because theorists and governments repeatedly insist that 20th century technical audacity is outdated, doesn't mean they're right.

For many, the turn to exploring infrastructure and skills in daily life is probably a matter of demystifying what has come to seem an alien world, administered by vast systems we neither understand nor control. In liberal societies, as geographers Maria Kaika and Erik Swyngedouw write, the latter "become abstractions, 'cease to be a product controlled by human beings', take a 'phantom like objectivity and lead their own lives'."[292] The "magic" of government, after all, is that it works in such a way that its ad hoc and deeply grounded nature is rendered invisible, such that it can appear natural and eternal, an abstract power in the eyes of the governed. But nothing is all powerful, nor eternal. And certainly not infrastructure or liberal life. Far from an abstract plan or the seamless enacted intention of political actors alone, techniques of government arise, rather, in response to a crisis, through efforts to govern this or that situation.

Only subsequently do these different strategies appear in retrospect to be part of a coherent rationality. Instead of subjects of powers greater than and beyond us, taking this perspective makes us into a hacker or engineer (or at any rate, less dependent). Exploring infrastructure from this angle, we discover that it is not really anything so special. Nor is infrastructure limited to its governmental uses in the front or back loop. Instead infrastructure is nothing other than what Karl Marx and Friedrich Engels called the means of life, the way people organize themselves for their existence.[293] As Angela Mitropoulos puts it, "infrastructure, after all, is about how worlds are made, how forms of life are sustained and made viable ... the undercommons, the weave."[294] In so far as it is almost always designed in response to local conditions, in order to project beyond or achieve security within them, all infrastructure is hubristic. Among the infinite techniques of human existence, infrastructure is the material means for making our dreams and ways of life live.

What the current problematization of infrastructure shows quite clearly is that the concrete, material reconfiguration of life is on the table as a question not of the future but of the present. That so many ordinary people are now studying infrastructures and skills, whatever their orientation, alone may be indicative of a society that is ready to reinvent the world. There is nothing to wait for in this regard. As urban dwellers from New York to San Juan saw first-hand during hurricanes of recent years, just as much as participants in social movements and uprisings have seen in their efforts over recent years to launch revolutionary transformations, infrastructures are both our life support systems—systems to which we are attached and on which we are dependent—and serious obstacles for any serious transformation of the world. Infrastructural experiments need not seek a return to a time "before" front loop infrastructure nor its mere continuation, and neither still need they simply aim to survive its ruins or disasters, as if liberal life is all there will ever or could ever be. Rather we can use the flotsam and jetsam of the front loop—including its technical audacity—to create our own pathways in the here and now.

5 Use of the Body

Let the bodies hit the floor

The possibilities for transformation open to us in the back loop are not limited to the practical skills and tools needed to survive, and the reduction of human and nonhuman aspirations to such survival is one of the most negative consequences of the resilience regime. In the face of this reduction, it is vital to insist that there are many other valences to existence. Recall the image of Victoria Falls used by resilience proponent Johan Rockström in chapter 2 to illustrate the danger of inhabiting a back loop time. It is worth noting that, in reality, Victoria Falls is not only a wonder viewed from afar via telephoto lens but a popular extreme recreation destination. During certain times of the year, locals and international tourists alike crawl or cannonball in and let the currents carry them to the edge. There they bathe under a violent spray of rainbow-colored rain amidst a thunderous precipice of a 350-foot drop.[295] The point is not to glorify extremophilia: people have certainly died. But the waterfall rather reminds us of a crucial fact: from living and playing in extreme or changing environments, to improvising post-disaster, humans are not unexperienced, nor even always averse, to thresholds or "edge" situations, and, most importantly, our experience in or beyond perceived thresholds is not only survival-oriented.

Consider popular phenomena emerging in the last decade within the domain of health and fitness such as CrossFit, natural movement or mindfulness. In each of these, people of diverse backgrounds—veterans, bus drivers, company execs, skater teens—seek to hone the human body and mind, in the process both discovering what are considered inherent capabilities and redefining the limits of their potential. In CrossFit gyms around the world, to take one example, overweight doormen run sprints and drill bear crawls. Middle school teachers heave 185lb barbells off the ground in an explosive movement, racking the bar to their chest, then dipping and driving the bar overhead. Kids and grandparents together run laps around the block, interspersed with burpees and squats. One after another doorman, teacher, kid ring a cowbell, a sign that they've hit a weightlifting personal record. Amidst an epidemic of chronic disease, as CrossFit founder and "fitness renegade" Greg Glassman says, a lot of people want and need to become more than chair-sitters. Through the "looting of practical and theoretical stores across fitness and sport," new fitness movements are being created with experimental bodily regimens amidst repurposed, formerly disused industrial architectures.[296]

CrossFit emerged in response to what in the 2000s was a dominant model of commercial gym fitness, exemplified by Planet Fitness and other big-box gyms: mirror-lined walls; neat lines of small free weights and machines for movements targeting individual muscles in isolation, seeking high rep movements to chisel bodies primarily for aesthetics with little attention to function; and financially structured on the sale of as many cheap memberships as possible in the hopes that most members don't go the gym or even forget they joined in the first place. But equally CrossFit takes the form of a response to front loop conditions more broadly: sedentary wage work that destroys bodies and minds or obesity fostered by sugar and junk food industries. Here what ruling classes thought regarding labor discipline in the 19th century—that "the common people had to be kept at their desks and machines, lest they rise up against their betters"[297]—is no less true today. In response, many have a deep desire to shape themselves into something new.

In contrast to standard gym models, crossfitters follow constantly varied, high intensity functional movements—whole body movements that are also useful in everyday life—usually paired with a zone or paleo diet focused on "real" food like vegetables and meats. CrossFit is a branded exercise regime that follows an open source model: workouts are composed of intensely performed combinations of functional movements—for example, "Fight Gone Bad" is wall-ball shots, sumo deadlifts, box jumps, push presses, and rowing. A new "WOD" (workout of the day) is posted online every day for free, and performed by people worldwide, each of whom scale the WOD according to their own needs and intensity. The movements are designed to be functional, meaning they mimic those involved in other useful activities from carrying heavy groceries to an all-out fight to a high-speed chase. Think hauling giant tires on chains, throwing medicine balls against the wall, pulling one's body weight up above and over bars, as well as climbing ropes, Olympic weightlifting, and calisthenics. People open their own CrossFit gyms ("boxes") in back yards, strip mall storefronts, or often disused industrial warehouses—wherever possible. Each box has a similar set of minimal equipment: pull-up bar rig; barbells and weights; horse stall mats with their signature rubber scent; ropes attached to the ceiling; airbrushed or graffitied slogans ("live free or die;" "we are what we repeatedly do;" "no whining"); oriented around a central whiteboard, where the WOD is listed and each person tallies their score to compare and contrast with others.

Looting of practical and theoretical stores across fitness and sport

Gregg Glassman incorporated CrossFit in 2000. Glassman, a wingnut genius who looks terrible describes himself best: "I'm a rabid libertarian. You *make* me do something, if I'm already doing it, I'll stop doing it. Even if I thought it was a good idea and it's something I want to do. No I'm not gonna be told what to do."[298] Rather than leaving fitness entirely, Glassman literally put the forms around him to new use in ways that suited his needs and taste. In one of CrossFit's mythical origin stories, teenage, then-gymnast Glassman was in his family's garage, trying to develop an intense competition-like workout using a $19.95 110lb Ted Williams weight set from Sears, Roebuck and Company.[299] He followed the directions in Ted Williams' Guide to Weightlifting, but for all the lateral raises and curls he did, none of them gave him the intensity of feeling he sought. Through trial and error he realized that squatting to the floor with the bar, then standing up to raise it overhead, did. Thus he invented a movement now known as a "thruster." Added to that, pullups on the bar he had jammed into the garage doorway. He tried 21-15-9 reps of each. He vomited, and then ran across the street, collected his neighborhood friends and made them try it. Everyone vomited; everyone loved it. "A workout was born. It's just that simple. You just try something."[300] And so on, a similar process unfolded over the coming years, experimenting on himself and friends. A college dropout who attended half a dozen universities, Glassman skipped from part-time job to part-time job, to underpaid trainer gigs across the gym industry. After getting thrown out of all the commercial gyms in Santa Cruz, Glassman opened his own gym, started a newsletter, and in 2001 put up crossfit.com with WODs.[301] People around the world did the workouts and posted their scores on the website's forums. It was there, Glassman recalls, "that we broke free from trying to accommodate the prevailing norm, both physiological model as well as business model, and struck out and opened our own gym."[302]

In 2005 there were 13 CrossFit affiliated gyms, in 2018 there were an estimated 15,500 affiliates in 162 countries, more than the number of Starbucks in the US.[303] According to *Maxim*, a new box gets opened every two hours somewhere in the world. As of 2015, 115,000 people had been certified to coach, and an estimated 4 million people do CrossFit around the world.[304]

In some ways like the amphibious architects, crossfitters take existing social forms and infrastructure—the very things causing disease and suffering—and put them to new use. Instead of working with individual machines, as in most commercial gyms, CrossFit gyms repurpose items like large tires and lacrosse balls producing new pleasures and new body cultures. In this sense CrossFit "destitutes"[305] late capitalist or front loop apparatuses of fitness and the cultures and body norms they entail. And rather than championing formlessness, in CrossFit one can see the great value of creating one's own forms, what they give to life and moreover the possibilities this opens up. Indeed CrossFit has a very specific use and technique, and is in fact replicable anywhere. You have some bars, barbells, maybe old tires, weights, horse stall mats. And you have a set of learnable techniques for using these tools: pull-ups, muscle ups, Olympic lifting, etc.

Along with these basic workout forms, the model for entering and expanding the CrossFit universe is simple because of its low barrier to entry. Adopting the decentralized techniques of management which have come to dominant business since the 1970s as well as the anti-corporate tech company vibe,[306] to become an affiliate you pay a fee ($500-3000) to use the CrossFit name (many also crossfit in gyms without the name, the same way many farms grow organic vegetables without getting certified due to cost).[307] Unlike the American Council on Exercise (ACE) or other certifications, this allows many more people, if they're willing, to start something, build businesses, and lives. People start boxes in their backyards, parks, on the side of roads, and the typical setting of most larger gyms is in old warehouses or factories.

Every box takes on its own flavor and personality (to the point where you would probably want to avoid half the boxes out there). Says Glassman, "I wanted to preserve as much of the entrepreneurial spirit as I could. I let people project as much of their own taste and style and values into this thing as possible," he said. "To not be told what to wear, what to speak, what music to play, what time to unlock the doors ... We leave that alone, because that's not the things that are important to the brand. What's important to the brand is the physiology, the methodology. ... We control the method, but the rest is up to them."[308] Through both its structuring and flexibility, CrossFit has opened up a new plane of reality inhabited by different people in their own ways, with some leaving past jobs and life trajectories to take it on full time, some becoming financially successful—crossfitters starting new lines of business, Rogue, etc.—while others take large paycuts and give up work stability for a more satisfying life doing what they love on the pullup rig.

Independent living

Many are critical of CrossFit, for reasons including its seeming proximity to neoliberal organizational structures (e.g. promoting constant precarity and readiness[309]); tendency to cause injuries; supposed lack of inclusivity; focus on going as fast as possible through heavy lifts; and so on. Perhaps the most common criticism however is that CrossFit's high intensity workouts are not appropriate for an average person.[310] Countless takedown articles warn readers of the high rate of injuries sustained by crossfitters and caution against its high intensity format, alternately calling it crazy, extreme, unsafe, dangerous, etc., echoing and reinforcing a societal obsession with safety.[311] According to *The Journal of Strength and Conditioning Research*'s 2013 study, CrossFit has an injury rate of 3.1 per thousand hours of exercise, the same as weight lifting or triathlon training.[312]

In fact, CrossFit is most popular precisely among those with no weight lifting or fitness background. For many it is the exhilaration of an intense workout resulting in sweat-soaked shirts, ripped, bleeding calluses, the joy of pushing oneself to overcome one's fears, limits, weaknesses, the learning of a new skill or transformation of one's body, and the ability to help each other do so that participants love most. And the same could be said for the popular Tough Mudder and Spartan races. For some this is explained in terms of how competition mimics ancient human instincts of hunting currently dampened by desk society—for some, as Kyle Kubler puts it, "CrossFit serves as the primal, libidinal release for those who work drab, "corporate" jobs and an affirmation of "lifestyle" for those who work for start-ups"—while for others it is a way of making oneself better, both for oneself and for one's family who depends on them; for others it is just that feeling.[313]

CrossFit has generated a million-dollar industry, corny mediatized CrossFit Games™, and is incredibly popular among the Silicon Valley tech elite, as well as all kinds of white collar workers, ex-marines, army, ex-SEALS, firemen, etc. As a result the stereotype of CrossFit—incessantly repeated in thinkpieces and leftist imaginaries—is that it's dominated by wealthy or upper middle class tech or creative industry types, venture capitalists.[314] This is both true and false. Rich tech nerds certainly populate many boxes, especially in cities with thriving creative industries. But in the Joe Schmoe nobody gyms, outside of the orbit of Silicon Valley or Manhattan, something else takes place. Drop in at CrossFit Breed in Queens, New York. Men and women who spent their working day holding doors for the rich on the upper east side; driving city busses; doing admin work at Riker's

Figure 5.1: An evening WOD at CrossFit Breed, Queens, NY. Photo by Marta Zapardiel.

Island; selling chicken wings, crepes, or sneakers in various stores; cleaning the apartments of the wealthier; teaching college students, are out in booty shorts and sports bras, socks pulled up to their knees for weightlifting. A Puerto Rican woman in her mid-thirties from Brooklyn heaves herself over the bar. In daytime, she is a grade school teacher in an overcrowded department of education classroom, teaching reading through rap to her students. Here she has transformed her body to become a workout hero, power cleaning her body weight. One of the basic ideas behind CrossFit—in spite of its sometimes hefty price tag (roughly $100-$175 a month)—is that these capacities are not reserved for an elite few, but belong to everyone, indeed specifically the average person.

Ultimately the CrossFit "type" or filter is not reducible to a bank account balance or a job sector as some have argued but it does have a less tangible vibe and character. Crossfitters are often people trying to get out of their comfort zone. In many cases, CrossFit is the first introduction to organized fitness while for others it's the continuation and intensification of a lifelong interest in physical culture. The CrossFit type, which is in some cases deliberately cultivated, is also someone who doesn't whine about a workout but who welcomes it as a challenge, doesn't lie about their reps; is humble because often failing in front of others; gives high fives all around. Very often the subjectivity found in boxes has "a psychic peanut allergy to top-down authority." Equally it is someone who seeks to change or better their life, who thrives from competition or a challenge, but also values being part of a community of competitors and challenge-seekers. "Who" in this case is just a misspelling of "how": you are what you do, and how you do it.

Wodapalooza CrossFit Competition, 2018, Miami: In the adaptive division a man is doing muscle ups. He's 6ft tall and paralyzed from the waist down. After each rep, his entire body falls to the ground, crumpling. Every single time, he hoists himself back up to do a movement that the majority of people, even in good shape, can't do. Women and men, young and old, even a teenager, in wheelchairs, place pads on their laps, so they can do clean and jerks, struggling, to use their upper body strength to lift the bar above their heads without assistance. The crowd roars in respect and cameraderie.

At stake for many people involved in CrossFit is the need to take control of one's own life. Responding to a culture of decrepitude and individual disempowerment—chronic health problems and obesity, actively maintained by the sugar and fitness industries themselves, medical industries designed to treat symptoms never causes—CrossFit may be seen as a back loop response on the bodily level which takes on the source of suffering directly, devising techniques for leaving it and its bodies behind. As explained in its Level 1 (L1) training seminar, CrossFit's functional movements are "essential to …independent living and quality of life," which help people age and stay out of assisted living and nursing homes. "I can't live alone because I need help, when I can no longer do independent things on my own, sink to a state of decrepitude." So "should grandma do Fran? Yes, and she will avoid assisted living." As such functional fitness movements mimic those you might perform in the rest of your life as you interact with your environment. A pull up gives you the strength to literally pull yourself up, over a fence or up from a fall. A thruster mimics the movement of getting yourself out of a chair, picking up a child, or putting groceries away overhead. The goal of a higher level of functionality is seen not just as functional competence but physical *dominance*—over psychological and physical barriers.[315] Instead of the normal definition of health as absence of disease—where one could still be in a state of decrepitude, in need of assistance—crossfitters see health as a movement for moving from sickness to wellness to fitness, maintaining fitness across one's years, and as buffer against decrepitude across life. A basic tenet is that no one need be excluded from this: everyone is equal to play the game provided we find our own right intensity.

From people who've eaten Ho-Hos all their life trying out nutritional challenges and Paleo diets, to grandmothers getting in touch with their inexplicable needs and desires, finding a feeling that drives them, crossfitters also cut themselves off from what they perceive as harmful to them. Faced with a world of sugar, chairsitting, obesity, and disempowering attitudes, they turn away. Once again here what you have is a practice of living, which does not define itself in relation to a perception of the world as catastrophic, but rather defines itself by its own positivity, practice, and source of power. Lifting, metcon, gymnastics are all such sources and avenues. "Competition help me fix myself," as one coach explained. "Each box, each gym, is a lifeboat in what is a tsunami of chronic disease" as Glassman puts it.[316] (To prove this, he has a "whiteboard in his kitchen where he's scrawled dozens of math equations that he says prove CrossFit has made people lose '80 million fucking pounds of fat'."[317])

Across different their different goals, crossfitters adopt an experimental ethos. As one of the leaders of a L1 training seminar put it, "you are in your own laboratory." Rather than following a standardized program, in every box, scores of people are exploring how to optimize and improve their practice. What began as a workout done by a teenager in a garage won't stay the same through 15 years of mass practice, and CrossFit and other fitness regimes are evolving as practitioners add to their shape and texture. Seeking appropriate forms of movement therapy appropriate to CrossFit's specific exercises and injuries, the anti-establishment attitude comes in again. Instead of the normal response—you get injured, doctor prescribes painkillers and stopping workouts—the idea here is to fix one's own practices in order to be able to continue exercise independently into old age. For example, one of the main causes of injury in CrossFit comes from bad lifting technique at high intensity. To fix this, in 2010 Kelly Starrett, started posting one-a-day MobilityWOD videos to YouTube, shot in his garage with bad sound and lighting, and featuring quick movement therapies to teach proper positioning and mobility practices ("DB shoulders"—doucebag—is the term he gives for hunched shoulders that come from sitting too long). Today his books on the matter, *Becoming a Supple Leopard* and *Deskbound: Standing Up to a Sitting World* are typically integrated in the regimes and bookshelves of most CrossFit boxes. These MobilityWODs often employ the aforementioned Lacrosse ball as myofascial release tool. Lay on or press the ball into tight tissue, roll on it or press it in for fifteen minutes, releasing tight, compacted muscle tissue. "Everyone should be able to perform their own basic maintenance," says Starrett.[318]

There are many other negative things one can say about CrossFit, and this is especially so from inside the practice itself. The many problems Crossfit practitioners themselves identify include the fact that it has given birth to a widespread nerd-jock culture giving dweebs a fake sense of coolness. Many boxes are just trying to sell personal training packages (making them dirtier, more expensive versions of Crunch, albeit run by guys who found a lower entry bar than standard gyms). Or, the fact that CrossFit's omnipresent coaches model is extremely annoying, with guys who paid $1000 to get an Level 1 trainer shirt thinking they can and should tell everyone what to do. The terror of community, the terror of coaches, the list goes on. All these issues stand in contrast with the antiauthoritarian spirit which, as already discussed, simultaneously pervades the world of CrossFit. Evidence of the way even a founder's own trajectories take on a life of their own, recently Glassman himself has not only gone to war against Big Sugar and the National Strength and Conditioning Association, but even his own brand, recently revolting against the CrossFit Games and redirecting funds and attention to local boxes and CrossFit fitness fundamentals.[319] Surely one could go on to write a book discussing the many issues with CrossFit. But the point here is only to remark on its existence as a phenomenon, emerging within the contours of back loop upheaval and reconfiguration just as much as "environmental" matters.

Reinventing mind and body

Many speak of CrossFit in the singular, but it is something being continuously modulated and created by millions of people practicing daily in their local box. I. From its personality to the WODs being invented each day—"we built this box," a slogan common on t-shirts—is also a deeper statement of what CrossFit is and is becoming.

CrossFit is unique but not alone. At any given moment, thousands if not tens of thousands of people worldwide are engaged in challenges such as the 21 day Better Human Challenge organized by End of Three, a fitness site and podcast, in which new practices like memorization, mental awareness, and cold showers are engaged as means of obtaining better mindfulness and mental equilibrium.[320] As the old models give way, new models are being forged. Finding themselves scrolling mindlessly, faced with what they perceive as the laziness or lack of care of their generation and our society, many are borrowing Stoic techniques developed in ancient Greece but learned from Instagram, to meditate, develop mindfulness, and master their passions.[321] "I think Stoicism is the answer to the problem of millennials," as one CrossFit coach, Prabesh Gurung, who is 23, put it. As Ido Portal, world famous movement practitioner explains, when we engage in these practices, "we go into this place, where what seemed impossible becomes possible, break through the fear of mind."[322]

In many ways these recent fitness models are evolving on a lineage of wild, unboxed American fitness (guys inventing bodybuilding reading magazines in 70s New Jersey; step instructors in 90s New York basements). Likewise, street workout, an evolving art of bar calisthenics, was developed in the early 2000s, around the same time as CrossFit, in the parks of New York and Moscow. Basic movements, from pull-ups to complex bar artistry such as stairs, were created by people sharing videos online, competing and trying to imitate and outdo each other (who's got the next new move)? The muscle-up, one of the emblematic movements involving pulling oneself up and over the bar, was arguably invented by the original barstarzz in the parks of the Bronx.[323]

Each of these body/mind practices intertwine, overlap, mutate in and with each other. This is also the case with CrossFit, which often functions as a portal into a new universe of practices and ideas. Crossfitters listen to YouTube videos on manifesting, or read Daily Stoic guides to Marcus Aurelius curated into tweet-sized summaries via daily emails with "2 minute reads" for "the 9 most important Stoic exercises" (memento mori: meditate on your mortality, amor fati, etc.[324]). Across these fields, every day millions of fitness practitioners share experience, tips, and links online through social media and message boards, literally constituting a new knowledge and skill set in progress.[325] If there is mass interest in Greek practices of the self from Stoic mindfulness to calisthenics and other physical arts, it is in so far as they provide one of many "stores" of practice now being drawn on as practitioners in garages, boxes, message boards and gyms constitute new arts of the self proper to our current back loop environment.

These practices do not fall within the traditional domain of what one would probably imagine as "Anthropocene experiments," if by the latter we continue to index only "sustainability" designs and bureaucratic management. Yet this physical fitness-focused wave is perhaps one of the most widespread, and transformative, amongst the range of human efforts to recreate themselves in the back loop. The forms of fitness being tested out and developed are both a response to multiple perceived problems—the need to get in shape, desire for a more intense or meaningful life, the tsunami of chronic illness, to change and improve one's life—and at the same time the instantiation of new ways of life and new kinds of human bodies. Across all these milieus is "the story of ordinary people, just like you, who discovered just how extraordinary they could be," as author McDougall puts it, who take the perspective that "heroism isn't some mysterious inner virtue" but a "collection of skills" that all humans, grade school teacher and doorman alike, have the potential to master.[326]

6 Use of Our Soul

Just as much as its other characteristics discussed in this book, the back loop so far is dominated by fear, anxiety, hate, and idiotic polarity. As the frameworks that once provided coherence to Western societies fall apart causing pain and confusion, heightened by a toxic, nihilistic social environment of resentment—manifest in horrible arguments with coworkers or friends about "politics" and horrific wars waged through social media—confusion and aggression deepens as many immerse themselves in internet echo chambers. In the back loop, remaking bodies is as serious as figuring out fire and water, shelter and food. But again still other dimensions are now open to us. What are needed today are not only technical tools and infrastructures to live or survive in the back loop, but perhaps more than ever, new feelings, energies, human dispositions, and subjectivities able to breathe new life into the world. Forms of life and energies not based on fear or resentment. Voices and songs of peoples irreducible to the current banter of left/right that seeks to close down openings now present. Other intensities of existence and ways of feeling, including just feeling good. With this in mind this section shifts gears once more, to explore yet another contemporary creating his own style of existence within his specific back loop conditions.

Roots revival?

Many describe Jamar McNaughton, a 26-year-old reggae singer popularly known as Chronixx, as the leader of Jamaica's roots revival, an island-wide return to reggae's hopeful 1970s golden age.[327] Watching the mini movie for Rastaman Ease Out, a song on his album Chronology, it's easy see why. In the *Rockers* homage/retro throwback, we follow then-21-year-old Chronixx, playing a young Rasta, smiling to the rising sun while birds sing sweet songs, as he leaves his doorstep of his humble home in the hills of the Blue Mountains on an errand to get peas for his girl. Strolling through winding neighborhood roads, giving high fives to friends in retro tracksuits against a zinc roof and cinder block backdrop, the young Rasta picks up cash and some ganja from a friend ("love the aroma!"). He gets into trouble with corrupt police drinking on duty, but as he quickly outruns the out of shape officers on a chase through Jamaica's lush countryside, the viewer is reminded why the Rastafarian Ital diet is not only healthy but a key survival tactic. In short, the feeling is light, and the questions seem simple. Roosters crow, banana trees unfold toward the sky, love from all, bless up.

Chronixx spent the first seventeen years of his life in Spanish Town, Jamaica. The island, which arose from the sea in the Miocene and is the third largest in the Caribbean, in recent centuries has been a "virtual laboratory" for the "unworlding" and "worlding" experiments that grounded the Anthropocene front loop.[328] Spanish Town was the Spanish and British capital of colonial government on the island, which the British used for almost two centuries as a slave colony, transporting one million enslaved Africans to work the sugar plantations, from which nearly half of sugar imported to Britain came, making Jamaica one of the most profitable colonies in the Empire.[329] Alongside plantation slavery, the British transformed a large swath of the island's biodiverse tropical forests and grasslands into monocrop plantations, draining the soil of minerals and nutrients, with mountains mined for limestone and bauxite.[330]

More recently in the twentieth century Jamaica was home to fierce anti-colonial experiments in building a unified, free nation for the "wretched of the Earth."[331] As Jamaican anthropologist David Scott recounts, the '70s were the island's "short decade of hope and expectation and longing. Whether you were a Rastafarian... or whether you were part of Michael Manley's democratic socialist People's National Party or the communist Worker's Party of Jamaica...you lived inside a surging momentum...for radical social change."[332] The dream of this generation, which included Rastafari and reggae at their height, was what Scott describes as a "postcolonial state which could impose a single standard of moral and civilizational value, a single idiom of rationality, and a horizon of ends toward which the population as a whole was obliged to head. E pluribus Unum: out of many, one..." Today however, "anticolonial utopias have gradually withered into postcolonial nightmares."[333] Less than one percent of the population owns the majority of Jamaica's land, and in Spanish Town, where Chronixx grew up, many do not have running water, and collect rain water in barrels for bathing, drinking, and cooking, often making ends meet through hustle economies and illegal water and electrical connections. Diabetes and hypertension are widespread, against a backdrop of gang warfare, beheadings, gun shots, and stabbings.[334] Disdain for the country's corrupt political and civic institutions is widespread, and moral authority is located more often in popular forms like dancehall (the latter seen as an index of crisis for supporters of the former), where artists like Shabba Ranks and Yellow Man supposedly trade reggae's upfull one love vibe and beachy one-two rhythms for hardcore computer beats, cocaine and violence.[335] Amidst this, the country's middle class, "now declining in moral authority, swings between urgent demands for a more no-nonsense and authoritarian policing and plaintively bewailing the collapse of civil society."[336] Those who once believed in the Leftist languages of emancipation, such as Scott, see Jamaica as a place of poverty not only economically but imaginatively, defined by the *"exhaustion* of the energies necessary to think it through afresh, *politically."*[337] Where just decades earlier the island looked to be heading toward a better, liberated future of postcolonial emancipation, concludes Scott, "one way of telling the story of contemporary Jamaica, increasingly volatile and frequently ungovernable, is to say that no one now has any confidence in that dream."[338]

Politics is like for 2,000 years ago

Born in 1992, Chronixx grew up at the tail end of this story. From his vantage point, the end of the old world, is probably not something coming, but a daily reality. So despite being labeled as the leader of a "1970s era reggae roots revival," in his own view, there is no going back. Not just because "what the ancient men" like "Peter Tosh and Bob Marley… used to experience …is a lot different from what we're experiencing now… five times as bad"[339]—social conditions are way bleaker now—but also because in his view going back is not desirable. I'm of *this* time, he says, part of our "Now Generation."[340] Paraphrasing words first spoken by Ethiopian Emperor Haile Selassie in 1963 which come across with new significance in the back loop, Chronixx says, ours is an "unprecedented" situation in human history, in which we face "new problems. Searching the pages of history for answers to these problems will only lead to a certain point and no further. Because these are brand new problems."[341]

Just as the questions are changing, so too are the modes of finding answers. For Chronixx answers won't be found in politics—"in Jamaica, there is no bright future for politics…politics in general is backward. Politics is like for 2,000 years ago…[when] Caesar was in a robe and like, "I am Caesar." Politics was cool. But now someone comes and says, "I am the prime minister of Jamaica." That's stupid. Rasta is not about that. Rasta is more dealing with love"[342]—nor are they to be found in the media—"think how many layers of makeup [they] put on… how many cameras… all the shebang, the helicopter and the overhead shots and all that shit. It's a show. But then you have people that are in the action, and they know exactly what's happening, they can *feel* it." Finally still, answers are not in the hands of a mystery god: as he recalls, he did church choir until he was 12, until, he says, "I decided... I didn't want to sing to a god in the sky. That's too far.""[343] Instead of looking elsewhere for answers or authenticity—up to the sky, back to a more authentic past, or beyond to transcendents that would justify or provide blueprints, authorities or government—Chronixx takes a different approach. For him, required today is neither a revival of the past nor a solution waiting off in the sky, but the generation of new cultures and aesthetics from the vast reservoir of resources found in the present and in one's own life. "We must look into ourselves, into the depth of our souls."[344] It's a matter, he suggests, of

> Becoming open to the sounds that exist within your consciousness. Sounds that you hear in your dreams, sounds that you hear in meditation, sounds that you hear in nature. This song that we're looking for, this song that we're trying to write, already exists. The birds already sing it before you. Thunder is the first bass, and the ocean is the first chimes.[345]

While there is a movement of younger Jamaicans taking up live instruments and reggae music again of which Chronixx is a part and supportive, he sees his project as something different. Instead of a return to roots, he says, call it "black experimental music." Weaving together diverse elements from his experience as a youth in what he calls the "rough training camp" of Spanish Town, colonial and postcolonial traditions, and "literally ... experiment[ing] with our soul,"[346] Chronixx is making an ethically and aesthetically powerful form of music but also *style* crafted from the landscapes and legacies of his own back loop. His training came from his father, dancehall legend Chronicle, who instructed Chronixx to use mop sticks and Guinness bottles as mics and to imagine palm trees as the audience. Because the reggae style Chronixx wanted to make was out of fashion, as a teenager he taught himself how to produce everything, spending countless days hanging out in recording studios, at home watching YouTube videos, practicing production software and hardware.[347]

Today he is one of Jamaica's most popular artists. On mixtapes like 2012's *Start a Fire*, produced with Major Lazer and recent full-length album *Chronology*, the beautiful but simple one-two rhythms of classic reggae get carried forward in a new register for new times, that refuses to accept the binary poles offered by contemporary Jamaican or global society. Reggae beats are infused with his daily practice of the Rastafari way of life, which emphasizes increasing one's experience of Upfullness and Livity, for example by transforming negative words into high words (i.e. understanding is made into overstanding) or replacing many letters with "I" to indicate the speaker's self-determination and presence; eating "Ital," a diet of things that are alive or natural; but most importantly, forwarding "one love" as a fundamental orientation to self and world, grounded in the spirit that runs through all but whose locus is in every being and must be expressed freely by each in their own ways.

Journalists usually describe contemporary reggae music as an antidote to the perceived harsher, more aggressive tone of dancehall that has dominated the Jamaican airwaves for several decades. But against moralizing exhortations to choose a side—democracy or dancehall, shorthand for bourgeois establishment or nihilist cannibal void, the latter of course being a construct made by the former—in Chronixx's songs dancehall is not rejected but equally drawn in as an ingredient handed down from his father, alongside myriad other sounds like nyabinghi, ska, and American hip hop. He further rejects the Jamaican ruling class's embrace of Rastafari, as a tourist brand and "safe" bulwark against dancehall culture. And in a characteristically complex stance, he's been quoted in the media labeling imprisoned dancehall artist Vybz Kartel's contribution to society as "cannibalism" but also supports the artist, characterizing him as "a fearless creative soul in this Earth" alongside whom he stands against a media for whom "to divide and rule is their only plan'."[348]

Some Rastas and reggae musicians, finding themselves unable to perceive poetry in the new, hearing in it only the lack of their old politics, label Chronixx a sellout and criticize him for not being "political" and thus not part of "the movement." Likewise such critics condemn Chronixx for suggesting we trust in *one's* most high self rather than *The* Most High conceived as an outside force, or for trying to redefine Rasta when that is "not allowed" because "there are rules." But for Chronixx, preserving an already-dead tradition for no reason is not the goal. Plus, as he sings,

> *People got expectations*
> *Will they love you? No guarantee*
> *People all need salvation*
> *Will they save you? No guarantee.*[349]

Along with his unique sound, Chronixx collaborates with photographers to select locations and styles for shoots, releases original visuals regularly, selecting and organizing rhythms, tones, colors. Most recently, for example, as spokesmodel for Gary Aspen's Adidas Spezial Spring-Summer 2017 collection, he created a clothing line mashing "the scent of the peanuts roasting, the sound system and the football"[350] at Prison Oval, a Spanish Town football stadium and dancehall at maximum security prison, together with Rasta, and with

British casual fashion developed on UK football terraces in the 1980s. In short, across diverse registers Chronixx forwards a powerful and unique style made through sound, image, and attitude, that flips the script on the modern colonial laboratory, transfiguring its parameters and puts them to new use. In his view this whole approach not a new thing. "Remember that black people in the Western world, our last names are "Smith" and "Brown" and "McIntosh." So we literally had to experiment with our soul to create music… Because, ina opposed to the people in West Africa, who grew up with thousands and thousands of years of musical practices, and the freedom to practice those ancient cultural music, we had to dig deep in our souls to find it." When interviewed by older radio hosts, the latter are often astonished by the young singer's perspective. "But how do you know all this?" New York radio station Hot 97's typically-arrogant Ebro in the Morning exclaimed. "And you're only 24?! "Yeah," smiles Chronixx, with a commanding attitude requiring no recognition or justification.

Wherever our most high leads

To transform the world does not only entail material infrastructures but also calls desperately for new kinds of human beings. This is especially essential today as the degradation of human subjectivity proceeds apace with Western civilization's collapse (the world's most powerful proclaiming themselves "bullied"; resilience's reduction of life to the terror of entanglement in catastrophic systems; the omnipresence of resentment, as both left and right search for someone to blame daily, etc.). The way real people live is in stark contrast to this hysteria. And in the face of its absurdity, many have already jumped ship, embarking on their own paths. As an example of someone exercising this possible relation to the back loop, Chronixx shows what it can mean to become one's own ground, and the power this holds to wreck the degraded images of liberal life repeated by the world's elites. Inhabiting the back loop can also be a matter of becoming new kinds of humans, with our own grounds. A self-assured ability to navigate and inhabit one's own reality, to serenely exude one's own meaning and style.

At first glance Chronixx's approach to doing so seems to resonate with recent critical thinking on how to live in the Anthropocene discussed in chapter 3. These imaginaries portray the Anthropocene via images of ruins, portraying devastated, broken worlds, and implicitly or explicitly forward visions of diminished life, where hubris is no longer allowed, ideas of future improvement said to be impossible, and creation and audacity denigrated as outdated artifacts of the

20th century. These visions characterize ours as a time of surviving—either more or less resiliently—amidst the limits of cramped spaces. It is almost impossible to imagine liberal commentators *not* heralding Chronixx as a voice of such resilience, rising out the ruins with incredible ingenuity. But isn't this patronizing? How many times must a singular existence be portrayed in such heroic survivor terms? What interests me, in contrast, is the way in which the style that Chronixx, as well the other experimenters I'm writing about, forward does *not* line up with the imaginaries animating earthbound and ruins thinking.

In contrast, Chronixx is just a youth coming from what to many academic theorists might appear to truly be a broken or "ruinous" place—and not just metaphorically but concretely—but who sees it differently. As one of his songs itself says, "me a victim? Never." It's not that he ignores the tribulations literally all around him: in fact he is more ruthless than most in identifying them (for example calling former President Barack Obama a "waste man," for not pardoning Marcus Garvey).[351] But rather than obsessing over or being dominated by darkness, he says, "we're placed in this situation for a reason. So don't reject it. Embrace it… If you can't find a reason, then you should leave… there's a beauty in it, and that's what we need to find out."[352] For Chronixx, instead of ruins and devastation, the way things are, "I accept it 100%," he says. By this he doesn't mean he tolerates it—"toleration which says yeah I hate you but I tolerate you"—but love, which is real acceptance, for all the universe.[353] Beauty isn't a matter of accepting conditions as they are given, and then "resiliently" finding beauty in "blasted landscapes" despite their "devastation," or merely surviving ruins of the old world. On the contrary: the "people," he sings, are tired of the "mediocre," and what's needed, he says, is to hubristically take the pieces and go *further*, to *transcend* the current situation.[354] Thus rather than continue Rasta and reggae traditions unchanged, he has internalized and mastered them, pushed them to evolve (so whereas Marley sang, "Don't worry about a thing / Cause every little thing is gonna be alright!" In Chronixx's words it is now "when the goin get tough the tough get going."). To Jamaicans who sit back and complain about music quality without making any, he calls on them to "step up their game"[355]—"If you can do it, why you don't just do it then? Music is not rocket science…"[356] "We are always so busy being the victims, we lose sight," he states. "We get freedoms and we don't use them. We use our freedom of speech to say, 'I'm a free man, free man,' but freedom is a thing your parents fought and died for, and now we use it to say that?"[357] But more broadly, he says, again echoing the Haile Selassie reference quoted earlier, "we must become something we have never been / And for which our education and experience and environment have ill-prepared us / We must become bigger than we have been / More courageous, greater in spirit, larger in outlook."

Rather than a "redemption song" to "put the pieces back together" into their correct state, one person's broken pieces become Chronixx's firm ground, on which he stands to project and create to go higher. Faced with a rifted reality where the old transcendents no longer work, like those who experiment with infrastructures discussed in chapter 4, Chronixx actually *disentangles* himself, becomes his own ground. Whether it's beer bottles as mics, palm trees as audience, ProTools or YouTube, colonial histories, fashion shoots, British football casual culture, the ocean's waves or the sun's heat—what Chronixx offers is a view of the possibilities present when we take up the world around us, without justification, moral or otherwise, to go *beyond* our given conditions. This is equally possible on an individual or shared basis. And it is something you can't always see, touch, or read. Rather than something one goes back to, in Chronixx's view roots—past legacies—Rastafari for example—as well as present conditions—Guinness bottles, music production tech, branding, and photography, the environment—are the ground one can stand on in this regard, from which we can weave and project to go higher—"wherever our most high leads." As he puts it, these are the tools to blast out of the worlds the youth are born into and into worlds of their own creation.

Mi dweet fi di love mi nuh dweet fi di likes

Regarding the Anthropocene, many repeat the refrain that "humans" were so powerful. Well, actually, not really. Across the front and back loop, from enclosures and reservations to wage work and deskilling, deindustrialization and Walmarts, the majority of the planet's population has been systematically tamed and disempowered as the structures of liberal life were rolled out and recalibrated. Much of the population has been separated from its capacities in deeply material ways, a process which continues apace today with environmental disasters (to understand the complexity of recent back loop dislocations in the America, for example, one must at least minimally begin by laying post-1970s trends in real wages and debt for the working class alongside climate change indicator charts[358]). In the contemporary context, while many older Americans are locked into a lonely Facebook vortex of vicious name calling and fake news, young people everywhere are besieged by media and government teaching hate and destruction, of each other and themselves. A Chronixx himself put it in an Instagram post the day rapper XXXTentacion was killed,

> Us the youths of today have found ourselves in a world where we are constantly being programmed to hate and destroy ourselves (by the media and all the other institutions that promotes the ideals of the self destructive societies aka "the world"). Our depression is often cultivated by the fact that The generations before us blame young children for a lot of the confusion and destruction that we are now facing on the surface of the Earth. As a result Many of us rebel against our very existence and the sacred body that our most high self has chosen to carry us into the greater dimensions of existence.[359]

What other kinds of life are possible today? Instead of being ruled by fear and anxiety what kind of people do we ourselves want to become? Reactive or active? How to become anti-fragile and loving, confident and free humans? Instead of more hatred of the self—a neoliberal ruling class ideology if there ever was one—living in, or better, beyond, the back loop calls desperately for new ways of loving and owning oneself, something no outside force can judge. Liberalism's catastrophes are not "our" fault, nor something we need continuously lament in guilt or "sense" in ever new ways. The back loop is just a situation from which to draw conclusions, the most obvious of which being that it is time to become something better. From this perspective humanity is in no way over but just continuing.

Against the hatred of the self seemingly synonymous with the Anthropocene Chronixx reminds us instead of the existence and need for other relations to the self, including those of love. Such a love is not based in overwrought statements about care, as one equally finds in neoliberal discourse. Nor is it something one pays lip service to while in reality casting judgments on others around them based on appearance or whatever other attributes. This other love can be found in uplifting oneself and others, pushing and crafting oneself further beyond the conditions into which we are thrown. A love that can exist in telling stories of the world and one's life, and allowing that life to be complex and contradictory, rather than unidimensional (a youth who in addition to being from Spanish Town ghetto went to church and learned music there, is influenced by British casual culture, who is both front and back loop at once). This love can accept and welcome one's own power, rather than falsely downplaying it (while simultaneously wielding it, as do many who verbally disavow power). Such a love and indeed power that seeks the world in order to explore it, is always the birth of something unique in this time. A power not achieved externally—via enforcement of punishments or condemnations on others, prosecution or call outs—but by becoming what one is and knowing what one loves. "The only force more powerful than hate is love. It's in a different class," explains Chronixx.[360]

7 Out of the Back Loop

Experimentation in unsafe operating space

Resilience and ruins politics tell us that we face a future without agency or imagination except perhaps that sufficient only to endure or envision disaster. Preppers, Chronixx, crossfitters, or amphibious fishermen remind us that such convictions are a fiction. Each of these stories stands alongside countless other back loop experiments that are redefining life in the back loop with their own tones, colors, and directions. Such experimenters do not seek a return to an imagined "before" the Anthropocene, its mere continuation, or simply surviving its ruins. That being said, they are not evidence of an "affirmative" or "good" Anthropocene.[361] There is no need to ontologize or celebrate them, as this or that favored subject promising redemption for all. They do not illustrate an imaginary "proper" way of living, do not service an existing ideology I wish to forward. They just are.

Each provides a window onto not only how the back loop is variously experienced—at heterogeneous scales, in contrasting tenors and geographies—but also the way that its questions and dislocations are taken up at diverse sites. I've told these different stories because they represent the efforts of ordinary people with "skin in the game" responding to the back loop in ways other than that predicted and proscribed for us by today's experts.[362] Truth and the future are being continuously reworked and recreated by people living in a range of scenarios who in turn draw on the resources of the past and present. As such the present is transformed and new possibilities open. Against dominant liberal binaries–left/right, white/black; us/them; management/catastrophe; survival/management—back loop experimenters are carving out other lanes and just driving in them. Where many fear the rift, they are comfortable on the brink, giving shape to new geographies. For them, just because old ways of being hubristic and living are passing away—thankfully, for many—does not mean no other hubris and no other living are possible. Such practices not only show the lie of resilience and the Anthropocene moral code described previously but also they offer a starting place for overcoming the many impasses of liberalism and the Anthropocene broadly speaking. That being said, they do not offer blueprints. Instead they offer ideas for how to leave the world of blueprints behind.

Perhaps these various examples—the bodily regimes of CrossFit, musical vibes of Chronixx, open source technical experiments—may

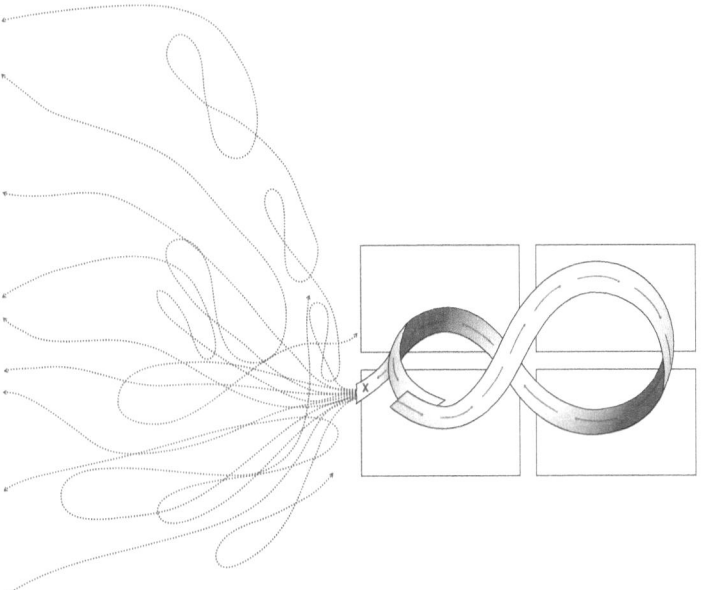

Figure 7.1: Adaptive cycle, modified to show potential for multitudinous exit pathways heading in unpredictable trajectories. Designed by Caroline Castro for this book, 2019.

appear random, disconnected practices. But amidst the back loop, such efforts are reclaiming and redefining the human being on Earth, forming the real substance of existence, the fabric of worlds being woven.[363] Or perhaps critics might say, "So what? It's nice their houses float, but I don't see how that's revolutionary." After all, based in recourse to this or that safe operating space—some heavenly paradise that will one day fall from sky and eliminate conflict; transcendent truths or political ideologies to orient meaningful life action; expert narratives to frame the latter correctly—politics has typically laid claim to a monopoly on transformation and historical action. But the idea that politics as the sole legitimate sphere of transformative, historical activity—especially given its current unimaginative, reactive state both left and right—would somehow survive the discombobulations of the back loop intact seems absurd.

Back loop experiments described in this book's second half take a different approach, and show another way of relating to the world, opening other modes of living untethered from the straightjacket of politics. There is something messianic to this approach in that, rather than take recourse to an external ground, it begins from one's own here and now. This approach is well-articulated by CrossFit affiliate owner Hutch Valentin, for whom inhabiting such a reality means that god is neither separate nor exterior from humans. Instead, following his views developed in the world of hip hop world as a Five Percenter of the Nation of Gods and Earths, a 1960s nonreligious offshoot of the Nation of Islam (NOI), god is something that humans can cultivate and develop through meditation, training in mental, spiritual and physical fitness. Rather than see the absence of transcendent grounds as synonymous with survival and ruins, this perspective says rather: we are our own ground, we are "god." Hutch's perspective itself emerges from his own life experiences and realities. While some Five Percenters see godliness as reserved for black men only, Hutch, who's Puerto Rican, has transfigured things into his own vision. He teaches his daughter and friends (who are many colors) that divinity is your true power—which is also our true power, which is the power of the universe, which is the power of creation of our universe and reality—and as such not only offers a useful perspective for us in the back loop but also yet another example of how people create their own realities. The idea isn't some megalomaniacal narcissism, but that each of us has the power to create our realities using our own hearts and hands. The "greatest weaponry available," as it has been put.[364]

Although unique, Hutch's perspective also resonates with the subterranean, often repressed but constantly reemerging, messianic tradition. The 15th-century Hussite rebellion in Bohemia, when the Taborites defeated the Holy Roman Empire, proclaimed their own Kingdom of Heaven on Earth and declared there would be no more servants nor masters; the 16th-century rise of Lurianic Kabbalah in the aftermath of the expulsion of the Jews from Spain; the heretical ideas of 17th-century Dutch philosopher Baruch Spinoza, who denied the existence of a transcendent God and said everything is one divine substance; Nietzsche's Zarathustra; and so on. Across place and time, from the Great Awakening's deeply democratic "religion of the heart" against the authority of reason and established church ministers, to those who meet today on the terrain of the back loop—what we might call a contemporary great awakening in its own right—there is a simple truth surviving across centuries and continents: the power of transformation belongs to regular, living people. We revise maps, we invent new practices, movements, and ways of living, in, with, and sometimes despite each other's unpredictable paths. This is a timeless reality, and a shift in thinking that is crucial for living in the back loop's "unsafe" operating space. Beyond transcendents—and freed from other authoritative yokes, be they political, intellectual, or cultural—each of us has the chance to become our own ground.

Free use

Having made this shift, rather than establishing rules for the whole world, the back loop experimentation described operates via a method of free "use." To "make 'use' of what's available" might sound like scraping things together, a form of salvage in order to survive within and despite the continued existence of (dominating) existing conditions.[365] But the "use" I'm interested in has a different meaning, and departs from treatments of the concept found across the works of Martin Heidegger, Michel Foucault, and Giorgio Agamben.[366] In the Heideggerian view, we are thrown into a world of people and infrastructure, geographies and climates, plants and histories, chatter and moods.[367] This world may not be of our choosing—so it can often seem like this order of things is inherent and no other possibilities are present—yet it is full of potential. The question is, how to become free in it? A liberated existence is found not by following and repeating the order as is, nor by celebrating it—to limit life to this, as does ruins politics, for example, is to live merely thrown and therefore to be precisely unfree—but by projecting ourselves within, over, and against such factical conditions. Instead of an untouchable power beyond us, the conditions around us instead appear from this perspective open to use and a new landscape of possibilities opens up. To be free is neither to escape nor live enslaved to factical conditions, but to become aware of them and to determine one's own relation to them.

More directly, when Michel Foucault turned to the Greek conception of use (*chresis*), it was to explore a concept of action opposed to moral code.[368] Whereas Christian moral doctrine views worldly matters such as forces of pleasure as bad or shameful, rooted in the Fall and human error, and therefore was concerned with rules of proper conduct and punishment—the establishment of a systematic moral code that would classify practices as either good or bad, acceptable or not, and that could govern each and all in a universal way— Foucault contrasted an approach focused on the way individuals form their lives as ethical subjects by making use of diverse techniques.[369] Techniques of "use," Foucault explained, were determined not by moral interdiction but by a number of strategic considerations of the user's specific situation—the time of year, the weather, one's social standing and age, in addition to one's training and ability. Are you hungry or tired? Is it cold or hot? Humid or dry? The user's personal status? Timing. And so on. In no case was "use" a matter of following or being subjected to an external law or rule. "Use," rather, was an art. In practicing it, one modulated and factored in a number of variables, adjusting oneself in kind. Equally, "use" in Agamben's recent treatment is neither prescribed nor simply arbitrary: it is determined by what is possible, and what is not, as well as when, how, and with whom.[370] Here we might come back to the fishermen at Old River Landing with their amphibious houses. While for theorists such as Timothy Morton Styrofoam is considered abstractly as emblematic of Anthropocene evil—a "hyperobject" beyond human comprehension—from Jacques or Buddy's pragmatic vantage point it can simply be deployed in other ways.[371]

For Chronixx a different range of techniques (melody, fashion, image, use of the soul) appear appropriate and their practice generative of quite different forms and subjectivities. In contrast to the utilitarian sound of "use," here it is useful to recall that Foucault's research on this matter was on the Greeks' art of making use of the pleasures (*chresis aphrodisiōn*): sensual and sexual, carnal, desiring, loving "acts, gestures, and contacts."[372] This work was related to his larger interest in the making of one's existence a "work of art," and new possibilities of achieving beauty and brilliance. The Greeks made use of the forces of aphrodisia, but people today may also make use of much more. It's not just, as one reading of Agamben might have it, that we "destitute" the conditions in which we find ourselves. For a richer and more feeling sense, the use of bodies between bodies may be enriched with writer Elizabeth Grosz's view of art as the way living beings respond to the "forces of the Earth": how we select and organize its rhythms, tones, colors, weights, textures into diverse forms, not in view of a predetermined end or any utility at all, but to create "sensation," to "intensify," "become expressive" and "become more."[373] For all the talk of survival, inhabiting the back loop also means discovering creative forms of pleasure in and with the Earth. In doing so as Chronixx describes in the case of music, it is "no longer reggae, or blues, or hip hop, or calypso, or soca. The music just becomes music. It becomes the sounds, it becomes the sonical value of emotions, the sonical value of colors. It becomes the audio aspect of your very soul."[374]

Through the "use" of environment, music, aesthetics, historical legacies and one's own body, amidst a world in freefall back loop experiments create their own forms for life, articulating a powerful alternative to the contemporary discourse of limits, survival and ruins. These diverse practices freely and confidently take hold of the pieces of a fragmenting civilization and put them to new use, not to survive, not out of fear, but in self-assured and creative efforts to remake and redefine life's texture in powerful ways.

The practices I've described are those of human beings as they freely take up and define their lives. They do so not according to a blueprint, but according to their own dreams, wishes, and needs. They make these real by use of their environments, legacies, and tools—"looted" as it were from the front loop—woven into new arrangements that suit them. Lacrosse balls become a myofascial tool. Refrigerators become fans. But just as equally, in many cases they continue the same use of old practices within these new arrangements. GPS data collection to track river levels. Wikipedia for communication, power cubes for energy. Rather than offering ever new forms of management, "use" is an affirmative way of living not about governing according to a rule, but ways of life themselves conceived as their own rule. After all, "life," as Agamben puts it, "is a form generated by living."[375] And the life you are living, is the life you are living.

None of this needs Agamben, Heidegger, or Foucault to be understood. After all, haven't I already argued at least fifty times against importing old frameworks that no longer serve us? Indeed—there is no necessary reason why Agamben or Foucault should be carried forward across the rift. As in *The Day After Tomorrow*, we may as well just throw our Nietzsche (or at the very least Heidegger!) books in the fireplace, if carting them along is going to hold us back from seeing and living the singularly new now present around us. However the perspective I take is that, as we explore our own paths, the back and front loop offer a wealth of resources from which we can draw as suits us. This includes Styrofoam and includes philosophy. Maybe Foucault, great thinker of the 20th century, still helps us comprehend our now. Maybe not. In other cases, perhaps it is our own experience that leads to the best insights. There is no one way, no tools that are pure and clean. Nor others that are off-limits. This is the approach I take with the back loop concept itself, deriving from foundations in resilience ecology and systems thinking with which I do not necessarily concur and moreover which ground the resilience regime's attempted extinguishing of our ability to imagine alternatives. It is easy to see how an uncritical use of resilience's conceptual model would risk importing this ethos and closing down, rather than opening, potential for imagination or liberation. My argument however is that, just as during a back loop soil and plants in a forest represent material for possible new configurations, so too are concepts such as the adaptive cycle available for new use, reconfiguration, and perhaps abandonment when they become un-useful.

Tools of freedom

In taking up such pragmatic orientation, people, to different degrees, become shapers of their own conditions of existence. This approach affirms the world, but it is *not* to be hostage to it in the way that Anthropocene ruins morality imagines. Rather than being *bound* to the Earth as if in chains—a situation where no progress, movement, or hubris is allowed—or simply detaching from it—for those who experiment in this way, the back loop becomes a matter of "practices of liberty," that is, the techniques one engages in not just to get free from a regime or set of relations, but to live freely, through the elaboration of autonomous and powerful forms of life.[376] These arts give us the means to transform not only ourselves but also our very modes of existence. The means for projecting with, as well as over and against, our environments, such practices give us the capacity to create our lives and to be active rather than passive in them.

In this light, we might reconsider "Prometheanism." Its meaning has become synonymous with humanity reaching the apex of insanity as the most powerful geological agent on Earth, imagining itself a force on par with a super volcano or an asteroid, all possible because this human species raised itself up as the rational orderer and center of reality. Welcome to the Epoch of Promethean Man, cue the Accelerationists, the Breakthrough Institute, the Singularity, the space colonies, and, on the other end of the spectrum, cue the critical world's almost univocal condemnation of human hubris as an outdated relic of the catastrophic 20th century.[377]

But if we return to the Caucasus where Zeus chained Prometheus and listen again to the imprisoned Titan's story, what we actually hear is how Prometheus had gone out among the humans and found them destitute, lacking all knowledge of the means of production, the *forms* that make life more than bare life. In hopes of liberating the humans, Prometheus went to Zeus and asked him to give the people fire. Zeus refused to share even a spark with them, as one tale tells it, "for if men had fire they might become strong and wise like us, and after a while they would drive us out of our kingdom. Besides, fire is a dangerous tool and they are too poor and ignorant to be trusted with it. It is better that we on Mount Olympus rule the world without threat so all can be happy."[378]

In response, not only did Prometheus expropriate fire from the selfish gods to share with humans, but he also shared with them many other tools—like architecture and music—arts necessary not only to human survival but to the infinite ways we elaborate the good life. A god who would not bow to the gods, the myth of Prometheus is about human hubris but also about the tools we use to wield our hubris. These tools give us the means to transform not only ourselves but also our very modes of existence. For millennia, humans have experimented with them, designing new ways to stay warm, better ways to feed themselves, how to move without being detected, how to prepare food, how to absorb and attenuate a variety of stressors from cold to combat injuries, how to hack, how to express beauty and meaning. Song. Tools are the myriad, infinite bridges that attach us to the world, that allow us to make use of it, and to give shape to and be shaped by it. By giving mortal humans these forms, Prometheus therefore gave them the capacity to create their own worlds. And through this gift a second one was given: the possibility of another future, one not determined in advance but a future that was an open question.

From this viewpoint, rather than an eternal image of the human being (which would then act on an equally preexisting reality)—which is either good or bad—there is instead a boundless range of tools just waiting to be invented and deployed.[379] These bridges between us and our environment represent the newest and oldest human fact. Throughout history when *homo sapiens* migrated from savannahs to tundra, when they tested out new tools for food, shelter, warmth, waste disposal, medicine, hunting according to new environments, or painted the caves of Sulawesi or Lascaux; in no case were they merely surviving, but rather throwing off the world as it was, raising themselves above and over it, as a way to be in it, to render it inhabitable. Tools enable us to go beyond given conditions, beyond merely surviving or enduring them, to take life in hand and shape it. And of course, in the process it is not the same human, which emerges unchanged, but rather one that is constituted in the use of tools themselves.

Autonomy

Living freely in the back loop may be neither to escape Earth nor live enslaved to it as a fundamental ontological condition but to determine one's own relation to it via free experimentation. The other back loop responses discussed in this book—resilience, Anthropocene imaginaries—also forefront experimentation as a mode of practice. But the experimentation within unsafe operating spaces under discussion in this chapter takes a different trajectory and ultimately seeks possession of its own form of life, articulating a powerful alternative to the contemporary discourse of limits, survival, and ruins. Free experimentation in "unsafe operating space" is not meant to signify a life lived in perpetual risk and vulnerability, nor does it venerate the abilities of vulnerable people to thrive amidst inevitable catastrophes. That is the resilience doctrine, neoliberalism's current, damaging and universalizing subjectivity which seeks to encompass all people in its embrace as vulnerable survivors. Free experimentation in unsafe operating space is rather something of a play on this resilience risk-obsession. It describes a way of living untethered from transcendents and past codes—hence "unsafe operating space"—that creates and embodies its own ground to stand on. That *disconnects* itself rather than tumbles around always already entangled in the world. And in so doing becomes what it is. Such a way of life does not wait for blueprints but, making use of what it needs—resources, objects, environments, legacies—develops its own path (hence "experimental"). Such a way of life takes up its own strategies, suited to its needs and conditions. Such strategies unfold on their own pace and tenor, and require no external justification. As such they concern autonomy: autonomy in one's own being, an autonomy that can be poetic, can concern imagination and images, can concern the material means of existence, as much as autonomy in defining one's own truths and trajectories. Such a way of living—which may equally be individual or shared—offers great possibilities for going beyond resilience and the Anthropocene moral code. Over and beyond this, free people living in this manner pose the possibility of practically and spiritually overturning the devastating system which resilience/ruins prop up and extend into eternity. Only free thinkers can make revolutions, after all. Elite voices continually belittle and power wash over the complexity that actually inheres in the majority of common peoples' lives, whether our sense of our selves, our relationships to each other, or our understanding of difference, right and wrong, and what is possible. Every one of the experiments in the second half of this book in one way or another speaks to this complexity and richness of life currently denied by such contemporary political discourse both left and right.

Each such experiment instantiates its own style and texture, richly specific to its context. The click of a tac go-bag snapping shut. Snow crunching under one's feet hiking to a wilderness skills camp in the Catskill mountains mid-winter. The scent of peanuts roasting outside Prison Oval. The crash of a barbell and smell of rubber horse mats in any CrossFit gym. The faint bobbing of Buddy Blalock's amphibious house in Old River Landing, the gentle warmth of its wood stove. As a writer I can only tell you about them because I have lived or been close to these stories in one way or another. There are infinite others that other people could tell. Each speaks their own meaning. Their styles of existence are infinite in number and nature. After all, human being has no proper meaning, form, or shape. We cannot, no matter from what angle, have our correct nature, tastes, or definition of the good life defined for us. Amidst a liberal order coming undone, between resilience discourse and Anthropocene theory, too often life is reduced to depictions of diminished survival and calcified into new imperative statements regarding all of humanity qua species. This degradation of imagination and human possibility is maybe the other half of the "extinction event" we are living through, to paraphrase journalist Andrew Sullivan. What is needed more than ever is experimentation with free thinking and other forms of living. After all, as geographer Nigel Clark reminds us: "it is not only the presence of life but the exceptional richness of its strata that makes our planet unique in the solar system."[380]

This richness can only be explored by living and expressing one's own gestures, ways of being, thinking, and creating. Such arts, which are possible in every aspect of living, are essential tools for reinventing what it means to be human. When people do this, as Gilles Deleuze once put it:

> It's a gust of the real in its pure state. It's the real that arrives and people don't understand that and so they say, "what is this?" Real people, or people in their reality, it [is] astounding. And just what [are] these people in their reality? It's a becoming.[381]

Instead of governing it, charting autonomous paths in the back loop implies existing and being situated in it as well as thinking the present situation through it. Rather a fate or crisis happening to us, to inhabit the back loop in this way is to dwell in and populate it, to take hold of and perhaps even take over as one does a host. It is possible even to belong to the back loop, to have one's own place within it, to be familiar, comfortable, and involved with it, rather than fighting against or living in fear of it. A habitual, everyday act of free creation and building: a peace within shifting terrain.

Plus ultra

Recall Latour's maxim for his earthbound, *plus intra* as opposed to *plus ultra*. *Plus ultra*, Latin for "further beyond," was the slogan adopted in 1516 by Holy Roman Emperor and King of Spain Charles V. The phrase is usually associated with global conquest, exploration, and Columbus's voyages to the New World but equally commonly used to connoting risk-taking and audacity (bizarrely the phrase is also inscribed on the walls of the coat room at Donald Trump's Florida retreat Mars-a-Lago).[382] *Plus ultra* was a reversal of the *ne plus ultra* of the Pillars of Hercules, which were a maritime marker in the ancient Greek world, marking the limits of the known world beyond which lay Atlantis, a land lost in the vast territory of the mysteries of the unknown. As in Latour's injunction to live within limits, the pillars were understood as marking the *ne plus ultra* ("no further beyond") of the known world and a warning that it was both impossible and unwise to go any further.[383]

Back loop experiments described in the second half of this book, however, entail a way of inhabiting the world founded in confident flight as much as gravity. Instead of simply moving downward, bound or tethered to the Earth, as in Latour's vision of *plus intra*, equally if not more present in the back loop are rich and complex terrestrial lifeways reminding us that, far from limited to survival, the hubristic question of recreating human modes of living in a fundamental sense is not only on the table but already being taken up by diverse people across the Earth. Rather than *plus intra*, for this growing class of experimenters, the motto of the back loop may actually be *plus ultra*, albeit transfiguring the term's modern sense and opening the much vaster range of possible operating spaces for human life on and with Earth as well as the strategies by which humans make use of their environments. *Plus ultra:* beyond what we were, beyond the old world. Which for many, it is not only possible but desirable to leave behind.

Through free use of tools from song to design humans orient themselves upward, not toward a higher transcendent power or off Earth, but, to paraphrase German philosopher Peter Sloterdijk, according to their own immanent criteria and vertical gradient.[384] This movement is not merely onwards to exist biologically, but upwards, beyond our conditions, by taking control of them and propelling oneself to ever-greater possibilities. In response to a problem, people in diverse environments are creating bridges between themselves and their world, in so doing elaborating and remodulating their own forms of life, which are not prisons but, we might say, laboratories in which the improbable becomes the real. These are tried out in unsafe operating spaces beyond environmental, social, or political thresholds, but rather than endured, living beyond thresholds becomes a provocation to new, audacious and improbable experiments, in the performance of which humans leave their base camps of ordinary existence and ascend "the mountains of improbability." And the experimenter him or herself, with each new step, builds herself as an ever-accumulating mountain. From this perspective, "the deep plays are those which are moved by the heights."[385]

But the Earth does not dream of you

I have thus far argued that the back loop presents an opportunity to reclaim and redefine human agency. This is neither simply reasserting the old hubris, nor forbidding the latter by all kinds of clever new means, but refers to new ways of being happy to be human, having the courage to simply follow one's own path and taste, and finding the best tools to do so. That said of course the back loop also promises reconfigurations and agencies outside of our control. While the Anthropocene thesis attributes so much destruction and domination to human action, as Nigel Clark writes repeatedly, so much of the Earth is still beyond us.[386] This is because the Earth's forces exceed our understanding, but also because, as in the Earth's molten interior or much of the ocean, we've never even been there. On one hand, Clark's work suggests great possibility: for all the human domination referenced by the Anthropocene, he reminds us "there remain a great many bio-geophysical avenues as yet unexplored or incompletely realized."[387] On the other hand, it suggests a fundamental unknown quality to engagements with Earth processes. They have desires and aims completely unrelated to ours. Moreover, other lifeforms actively use worlds we produce to construct worlds of their own. Inhabiting the back loop thus entails not only that we allow ourselves to see our environments as open to rearranging, but also as rich in their own right and capable of rearranging us, too. Theorists might like to imagine the Earth is always entangled with us, but argues Clark, much of the world is simply for itself, beyond and unknowable to us. "This is the domain," he writes, "beyond our control, our knowing or our adjudication, and therefore beyond political purchase in any conventional sense."[388]

This simple and indeed beautiful fact need not require that we disavow our own capacities or sink into to nihilistic meditations on doom. Faced with the complexity and power of our Earth's movements, forces, and colors, and the beauty, provocation, pain, and challenges they often bring, why not see: an incitement, a provocation, a gift, an unknown, a singular presence, anything but an enslavement. That the world is in many ways autonomous and unknowable to us—who would really say otherwise?—does not require we also say, all power to the nonhuman. We can explore the kinds of life possible amidst volatile inhuman forces, and in this way reinvent our *own* autonomy. As Clark writes, recognition that Earth is made of forces beyond human control "need not stop at melancholic musings on fragility or finitude, but can be taken as an incitement... the opening that propels us into previously unthinkable possibilities of knowing and doing."[389]

Such is the richness of inhabiting the back loop. The Earth pulls us down into it, at times rupturing our whole life. We suture. Build bridges. The Earth consists of valleys and mountains. Air, oceans, deserts, swamps, plains. The sun rises, peaks at midday, sets, it must go down. It rains for days with no reprieve, the sun hides behind clouds, streets flood. Storms, wind, snow. The passing shadows of clouds overhead. Animals or people hunting and digging. Creatures walk, crawl, or run across Earth's crust, tied to it by gravity's pull. Others soar and glide over and above the planet. We may love our environments, but also at times need to project against them. A living terrestriality is a journey: the serpent slithers across the Earth's crust, the eagle soars above. Flight need not be constant, and we rest as well. Embracing this multivalency can be a source of great strength and power, expressible in a million registers. Soft, hard, poetic, and warlike.

Amidst a process of social mayhem, this will look unexpected. Drawing a final resonance with geographer Jamie Lorimer's work on rewilding may add additional depth to the notion of back loop practice.[390] In contrast to most conservation efforts, which try to plan and manage nature, Lorimer notes that rewilding, though discursively reliant on images of untouched, pre-human nature, does not reproduce or save a previously existing version of nature. Rather, in practice these are "wild experiments" that produce new and unexpected configurations, generated through the interaction of multiple forms of life. The "labs" in which these configurations are generated are neither pristine nature nor a perfectly controlled space of a typical scientific laboratory. Rather rewilding's laboratories are the "inhabited and thus political landscapes and ecologies of the Anthropocene."[391] Geologic, metaphysical, sociopolitical strata alike make up the ground of the back loop—a paradoxically fractured, vanishing, flooding ground—on which we will, if we choose, create the new. A lab in which nothing is certain nor neat and clean. What is required is often making life live in environments that may be functionally extinct, transforming, very hot or underwater.[392] And more immediately, the tools we try may not work. Or they may break. The water might rise much faster than we expected. Or: what if it rises more slowly than projected? What if regimes and markets stabilize, and what we await is but the unremarkable slow repetition of present conditions, albeit outfitted with more drones and artificial intelligence?

Ultimately no one knows if experiments will succeed, or exactly what kind of life—human or otherwise—will emerge, in part because of the inherited contexts and conditions that are involved. In short inhabiting the back loop will be a wild but also a "speculative" experiment, recreating and belonging intimately to the world.[393]

The future will be full of mutations and creations unlike what we've seen before. No one can say what is coming. Not knowing—the unknown—does not mean we have to accept the powerlessness of an incomplete life. Not knowing where we are going—or not going where liberal modernity's own story said we should go—is not necessarily an experience of terror, nor need it define us as precarious. It means rather that life is not a finished fact but instead a question, and that this is especially so today as transcendents continue to wither. As such, old frameworks should not be mobilized thoughtlessly to understand new, singular realities. Likewise it means we should begin from the real, and not abstractions of it. Finally it means that we can welcome the presence of each moment, rather than shoving it into a box from the past. All this isn't such a big deal. It just means you can try anything.

Into the unknown

Resilience, survival amidst the ruins of a broken world, free experimentation: these are all legitimate ways of responding to the back loop. In this book I am making a claim for the third possibility, but it is by no means the final answer. It is just the beginning. This is a process of exploration and of creativity. Of finding one's own way.

Many responses seek to either freeze this ongoing process or return life to "stable" grounds, even the new stable grounds of precarity, flux, and entanglement, to which the Anthropocene is said to deliver us. But contrary to continual statements regarding the "obligations" posed to us by the Anthropocene—antihumanism, entanglement, despair—neither the Anthropocene nor its back loop "tell" us anything. If anything the Anthropocene back loop is only the shorthand for an historical moment, one marked by profound transformation. As such it can only taken up, explored, inhabited or not. There is no one way to inhabit the Earth or to move on it, and no authority with the right to proclaim it. There are only the infinite experiments through which we enact our selves and worlds.

Such experimentation of course is haunted by the looming threat of liberal regimes which, hostile to the emergence of other definitions of life, seek either to incorporate or eliminate them. What would happen, for example, if and when government-mandated evacuation programs are instituted in flooded regions where residents have chosen to stay? Answering such questions necessarily will lead those concerned into a host of ethical considerations unique to each case.

In any event, back loop experiments need not justify themselves according to external yardsticks or radical accounting measures, which accord value to human activity—and one could argue life and existence more broadly—only if they can be instrumentalized or put to work by one or another political ideology or movement. That the answers to living in the Anthropocene can already be found in a series of a prioris cobbled together from imploding political frameworks should be suspect. Insisting that the forms and possibilities of the future subsume themselves to such languages, nay governance, should be doubly so.

The sun has set on such thinking, and back loop experiments simply move on an altogether other plane. That is what makes them interesting, and so potentially generative of alternative lifeways and possibilities. Rather than try to maintain old safe operating space or freeze a process already in motion, what is required is a time of exploration, in which we allow ourselves and others to find out what more we can become and where else we can go.

My suggestion that we are living in the back loop of the Anthropocene means that we have *already* crossed various tipping points, but that in doing so everything from social practices, technologies, and truth to plants, animals, and places have become shaken out of their normal frameworks. This shifting ground is at the same time our common ground. Rather than leaving us without meaning or agency, we have the chance to finally leave the safe operating space of sovereign guidelines and follow our own paths. As such, the back loop joins us to one another while immediately throwing us back to our selves, our real worlds with their dreams, friends, geographies, needs, and visions, which now provide the only real yardstick we need. The tools we require, what will give our lives meaning and power, will be found there, an interior place from which we are thrown outside of ourselves again, to a new plane of reality where we find out what and who resonates with us and what does not. There, and only there, is truly political existence possible, an existence that is not final nor only peaceful but an open field of love and hate, tragedy and heroism, defeat and surprise, ridiculous failure and success, indescribable valleys and peaks.

The back loop will be a battlefield: a war over words, ways of life, and visions for the future, as well as our ability to imagine or build anything else. While critical theorists tell us hubris is no longer acceptable and to instead celebrate the life of things or a world without us—bolstering a regime of governance which tells much the same story—diverse factions of the world's elite are getting organized to thrive amidst a civilization in freefall. Sea walls to protect Wall Street; luxury bunkers in New Zealand; Russia's global warming future strategies; Google's infrastructure on the moon and geoengineering Mars. From each of these and so many other angles, the imaginativeness called for by a back loop is channeled towards management, dimming down the horizon of possibility and extinguishing our ability to imagine alternatives. Faced with this, we have to fight for our ability to imagine, to dream and to create other worlds, but also to define their terms. Not just food, shelter, water, but how might we reimagine life, beauty, excellence, war, peace, security? What will the human, posthuman, or posthumous be? Inhabiting the back loop will be about figuring out what kind of life we want to make live, what kind of life is worth living. Instead of trying to come up with "what's next" or manage our end, we can explore the possibilities already present here.

Big picture concepts such as the Anthropocene often have a homogenizing effect, as if there could be a single "we" of humanity, as if "we" would all be equal in formation of the Anthropocene, or the experience and participation in its back loop. However, my aim has been to show that the back loop is a heuristic which helps us see that what is happening now is the formation of innumerable worlds and "we"s, the plurality of life which only the front loop ever thought to cover over. While we may conceive our time broadly at the epochal or civilization scale as a back loop, there are equally billions of back loops, and with many realities unfolding in them.[394]

Here think for better or worse of searches being carried out by diverse people for new practices and ways of explaining their life experience, from new and old fundamentalisms, fitness and cultural movements, to "new tribalisms" both IRL and "memetic" digital tribes.[395] With the loss of credibility of governmental regimes as well as universal sciences, with media increasingly seen as promoting biased stories and not the objective facts of an objectively existing world, here we may also file the rise of "post-truth" and its critique of expertise and media objectivity.[396] Such have democratic and anti-democratic, left and right, "pro-" and "anti-" capitalist expressions. Others do not fit in any of these boxes. What, one might add more darkly, of the array of lifeways emerging in America's "hinterland," such as those associated with the conservative Patriot Movement or Oath Keepers, which seek creation of local power bases and autonomy from the federal government in response to social and economic crisis?[397] Each of these examples for better or worse speaks to worlds which exist and are practiced today in direct response to the social, economic, and political dislocations of the back loop, but they do not conform to the images of entangled, anti-human life said normatively to constitute it.

Just as Jan Zalasiewicz reminds us that there is not one Earth but rather "different Earths that have succeeded each other in time," so too can we see that there is not one world but many ways of living, some of which may coexist while others may lead to ontological conflict and struggles for what Kilcullen calls "competitive control."[398]

Breakaway civilizations, breakaway forms of life, great tides of confusion and searching. Contemporary experiments reveal a variegated landscape of practitioners already inhabiting the back loop: malevolent back loop practices that seek to capitalize on its disruption, efforts to discipline it, but also less overdetermined ways of dwelling and creating in it. Each follow their own forms of knowledge, practices of truth and imagination, technologies of power and infrastructure, as well as ways of constituting oneself as a subject on the basis of these practices. Some of these paths taken by people inhabiting the back loop may appear insane, illegitimate, or offensive to the modern mind as well as the emerging Anthropocene mind. This illegibility however does not make them any less real. Inhabiting the back loop, as I have proposed, will neither be experienced homogeneously by people across place and time, nor lead to a cycling back into the previous infinity loop. As depicted in Figure 7.1, the back loop is not a single path for a single humanity, but innumerable forms of both.

This process is not something to lament but to embrace and encourage.

As for how to do so, we might find ideas in the following words of Tekarontake aka Paul Delaronde, known for his role along with other members of the Mohawk Warrior Society in the 1974 seizure of abandoned camp for the wealthy at Moss Lake in upstate New York's Adirondacks, leading to a three-year standoff with state and federal police, and the eventual reclamation of 600 acres of autonomous Mohawk Territory of Ganienkeh.[399] Telling stories in a longhouse in Akwesasne during the winter of 2015 he explained:

> Somewhere along the line, we became a people of faith, hoping we'd be saved. A lot of people had faith, but they never gave themselves the ability to implement it. But we are beginning to think again. Everyone is supposed to kiss the ring of the Pope, the general, the police, to show their allegiance. But the Christian god is just an invention of men to control other men. And what we're told are "laws" are just the opinions of individuals who want to control other individuals. The only law is the way (of a people, a path, a way)... No one—no man, no police force, no government—can tell us what way to be, or what way the water goes, or the bird or the bees. Stop hoping, waiting. You want to believe in something? Believe in the sun, believe in these trees, believe in this air, believe in us, and yourself, because that is real. Let that be your beginning, and let that be your end. There are many paths, for example we have our "canoe" and

they have their "ship," and we will never try to steer their ship, and in return they must never try to steer our canoe. But if their ship starts to capsize, we will help them get back so we can sail together.... We never want the world to be one red, all red roses.... Governments are a plague on the land and the people... A land is free, a people are free, in their way of life, with nature, with each other, using, remembering. That is how we will never be submitted, or become subjects of rules and regulations made up by governments, police, or courts who think they are the great "father" of all of us... Really we have to ask: are we our own masters? In our existence? How much did we listen? How much are we prepared to do? We might be having a tough time, but we're not stupid... There's nothing we can't do. We either will or we won't. We're asking questions of things we know the answers to.[400]

Coda

Like a Shell futurologist, one can imagine multiple disastrous futures for coastal cities like Miami. Will it become a southern Super Venice, a la Kim Stanley Robinson's New York of *2140*, a watery playground for the rich and capital speculation: Zaha Hadid-designed high-rise condominiums retrofitted so as to actually function with permanently elevated sea levels; floating tourist traps on Ocean Boulevard hocking $35 margaritas for what remains of the world's non-inundated middle classes; a motley and still surviving working class that ferries in from the overstuffed, substandard housing complexes where they live a kind of managerial socialism of long lines, board meetings, and just-in-time and rarely adequate dinners? Perhaps the hard realism of Paolo Bacigalupi's *The Water Knife* is more apt. The "haves" live in a series of Amillarah Private Island arcologies, closed-loop glass-domed living systems with luxury malls, fine eateries, and augmented reality advertisements in their centers and ringed by air-conditioned penthouses whose waste water is filtered into the loop, while the "have-nots" are clustered in camps along the new coast lines where they have the new Dust Bowl refugees (the "Floridians") gathering around pay-to-drink Red Cross water dispensers as they try to fend off the latest Chikungunya or Zika. Or imagine a super Katrina resulting in something a little more *Odds Against Tomorrow*: Miami a post-flood "dead zone" abandoned by government and left to rewild, reclaimed by pythons and alligators and scores of individualists with camping packs on their backs starting from scratch, spending their days transforming soggy banks into their dream apartments and building ramshackle boat homes amidst growing marshland. Meanwhile, more refugee camps in the background. Each of these visions undoubtedly carries an element of truth, but only if we allow it.

For quite some time, governments deployed a powerful narrative of progress: development, growth, and endless improvement. Now, many resilience advocates substitute our ability to shape the future with an "oops, we actually can't" survivalism that is hidden behind pristine architectural renderings. We may live in a world that increasingly tells us there's no more dreaming (except about space), but most people don't live like that. For many, our dreams aren't about the future or the end; they are about the possibilities opened right now. Taking up the challenge of the back loop is fundamentally a wager of the present, and it will only be met through a combination of adaption, reinvigoration, and a radical shedding of obsolete technical, social, and mental systems. To accomplish this, we'll need to make the unlikeliest of combinations between practitioners of all kinds. Perhaps the hard hats need to meet the hackers, and the engineers the ecologists, and the nurses need to meet the artists, bus drivers, teachers, and mechanics. We are already all here.

At the end of the day no one knows what's going to happen.

Maybe the rich will leave the flooded cities behind and we'll get to keep their boats, to live with the water like people have always done. I hope we will draw the new maps of a people who poetically make new out of old, who are not slaves to the suffocating idea of life as suffering/survival/interlinkage. The new standard for the back loop is whether you chose to participate in this process or not. From there, there are many ways to play, and that is fine.

As Jean Baudrillard once said, perhaps the main rule is that the game continues.[401] It won't be without paradox and contradiction, hardship and heartache. But that's fine too. There is no blueprint. Into the unknown.

Notes

1. Brian Walker and David Salt, *Resilience Thinking: Sustaining People and Ecosystems in a Changing World* (Washington, D.C.: Island Press, 2006), 1.

2. Many recent academic works focus on experimentation. For an effort to theorize experimentation from an autonomist political perspective drawing from social movements see Dimitris Papadopoulos, *Experimental Practice: Technoscience, Alterontologies, and More-Than-Social Movements* (Durham, NC: Duke University Press, 2018). On the turn to experimentation in climate governance, see Bruno Turnheim, Paula Kivimaa, and Frans Berkhout (Eds.), *Innovating Climate Governance: Moving Beyond Experiments* (Cambridge: Cambridge University Press, 2018); Matthew J. Hoffmann, *Climate Governance at the Crossroads: Experimenting with a Global Response after Kyoto* (Oxford: Oxford University Press, 2011); Harriet Bulkeley, Vanesa Castán Broto, and Gareth A.S. Edwards, *An Urban Politics of Climate Change: Experimentation and the Governing of Socio-Technical Transitions* (London: Routledge, 2015); Joshua Evans, "Trials and tribulations: Conceptualizing the city through/as urban experimentation," *Geography Compass*, 10, no. 10 (2016): 429–443. On historical forms of experimentation amidst abrupt climate change, see Nigel Clark, "Volatile Worlds, Vulnerable Bodies Confronting Abrupt Climate Change," *Theory, Culture & Society* 27, no. 2-3 (2010): 31-53. Doi: 10.1177/0263276409356000.

3. Michel Foucault, "Truth and Power," in *The Foucault Reader*, ed. Paul Rabinow (New York, Pantheon Books, 1984), 51-75.

4. Owen Gaffney and Will Steffen, "The Anthropocene Equation," *The Anthropocene Review*, 4, 1 (2017): 53–61. Paul Crutzen and Eugene Stoermer, "Have we entered the 'Anthropocene'?" *International Geosphere-Biosphere Programme (IGBP) Newsletter*, 41 (2000): 17–18.

5. Will Steffen, Johan Rockström, Katherine Richardson, Timothy M. Lenton, Carl Folke, Diana Liverman, Colin P. Summerhayes, Anthony D. Barnosky, Sarah E. Cornell, Michel Crucifix, Jonathan F. Donges, Ingo Fetzer, Steven J. Lade, Marten Scheffer, Ricarda Winkelmann, Hans Joachim Schellnhuber, "Trajectories of the Earth System in the Anthropocene," *Proceedings of the National Academy of Sciences*, 115, 33 (2018): 8252-8259.

6. Tipping point from Owen Gaffney, "Walking the Anthropocene," *National Geographic*, 2015, March 16. https://www.nationalgeographic.org/projects/out-of-eden-walk/blogs/lab-talk/2015-03-walking-anthropocene/. "Planetary terra incognita" from Will Steffen, Paul J. Crutzen, John R. McNeil, "The Anthropocene: Are Humans

Now Overwhelming the Great Forces of Nature?" *AMBIO: A Journal of the Human Environment,* 36, 8 (2007): 614.

7 Stephanie Wakefield, "Man in the Anthropocene (As Portrayed by the Film Gravity)," *May,* 13 (2014). http://www.mayrevue.com/en/lhomme-de-lanthropocene-tel-que-depeint-dans-le-film-gravity/.

8 Jan Zalasiewicz, Will Steffen, Reinhold Leinfelder, Mark Williams and Colin Waters, "Petrifying Earth Process: the stratigraphic imprint of key Earth System parameters in the Anthropocene," *Theory, Culture & Society* 34 (2017): 98.

9 Peter Limberg and Conor Barnes, "Memetic Tribes and Culture War 2.0," *Medium* (2018), https://medium.com/intellectual-explorers-club/memetic-tribes-and-culture-war-2-0-14705c43f6bb.

10 Gean Moreno, "Glitch-People," *Affidavit,* November 21, 2016. https://www.affidavit.art/articles/glitch-people

11 Stephanie Wakefield and Bruce Braun, "Oystertecture: Infrastructure, Profanation and the Sacred Figure of the Human," in *Infrastructure, Environment, and Life in the Anthropocene,* ed, Kregg Hetherington (Durham, Duke University Press, 2019).

12 Jane Bennett, *Vibrant Matter: A Political Ecology of Things* (Durham: Duke University Press, 2007); Alan Weisman, *The World Without Us* (New York: Thomas Dunne Books/St. Martin's Press, 2012).

13 C.S. Holling, "Understanding the Complexity of Economic, Ecological, and Social Systems," *Ecosystems,* 4, 5 (2001): 390–405.

14 C.S. Holling, "Resilience and Stability of Ecological Systems," *Annual Review of Ecology and Systematics,* 4 (1973): 1–23.

15 There is a large body of critical literature around resilience thinking and its foundations in systems thinking, cybernetics, and neoliberal economics. See, e.g., Bruce Braun, "A New Urban Dispositif? Governing life in an age of climate change," *Environment and Planning D: Society and Space,* 32, no. 1 (2014): 49–64. Doi: 10.1068/d4313; David Chandler, *Resilience: The Governance of complexity* (Abingdon: Routledge, 2014); Brad Evans and Julian Reid, *Resilient Life: The Art of Living Dangerously* (Cambridge, UK: Polity., 2014); Sara Holiday Nelson, "Resilience and the Neoliberal Counter-Revolution: from ecologies of control to production of the common," *Resilience: International Policies, Practices and Discourses* 2, no. 1 (2014):1-17; Kevin Grove, *Resilience* (Abingdon: Routledge, 2018). As will be clear in the following chapter I take a similarly critical stance toward resilience, in particular its projection of life as cybernetic systems onto all of reality, and as an ecocybernetic mode of governing the back loop. See Stephanie Wakefield and Bruce Braun, "Governing the Resilient City," *Environment and Planning*

D: Society and Space, 32, no. 1 (2014): 4-11. Doi: 10.1068/d3201int. Thus it bears noting here that while I am borrowing the back loop concept from resilience thinkers, I do not import their other epistemological or governmental assumptions with it. Such I will argue is a useful ethos for living in a back loop, the ability to freely make use of concepts and tools. Lance H. Gunderson and C.S. Holling, *Panarchy: Understanding Transformations in Systems of Humans and Nature* (Washington, DC: Island Press, 2002).

16 Ibid.

17 Ibid.

18 Brian D. Fath, Carly A. Dean, and Harald Katzmair, "Navigating the Adaptive Cycle: An Approach to Managing the Resilience of Social Systems," *Ecology and Society*, 20, no. 2 (2015): 24. doi:10.5751/ES-07467-200224.

19 Gunderson and Holling, *Panarchy*, 33.

20 Holling, "Resilience and Stability."

21 Holling, "Understanding the Complexity."

22 Lance Gunderson. Interview by Stephanie Wakefield. Florida International University, Miami, Florida, February 8, 2019.

23 Lance Gunderson, C.S. Holling, and Stephen Light, *Barriers and Bridges to the Renewal of Ecosystems and Institutions* (New York: Columbia University Press, 1995).

24 C.S. Holling, "From Complex Regions to Complex Worlds," *Ecology and Society*, 9, 1 (2004): 11. http://www.ecologyandsociety.org/vol9/iss1/art11.

25 Thomas Homer-Dixon, *The Upside of Down: Catastrophe, Creativity, and the Renewal of Civilization* (Washington, D.C.: Island Press, 2006), 228.

26 Holling, "Understanding the Complexity," 398.

27 C.S. Holling, "Resilience and Life in the Arctic," *Resilience Science*, April 5, 2011. http://rs.resalliance.org/2011/04/05/.resilience-and-life-in-the-arctic/

28 Gunderson and Holling, *Panarchy*.

29 Holling, "From Complex Regions."

30 Ibid, 4.

31 For diverse treatments of adaptive cycle, see, e.g., Fikret Berkes, Johan Colding, Carl Folke, eds., *Navigating Social-Ecological Systems: Building Resilience for Complexity and Change* (Cambridge:

Cambridge University Press, 2003); Fath, Dean, and Katzmair, 2015; Brian Walker and David Salt, *Resilience Practice* (Washington, DC: Island Press, 2012). On multi scalar panarchies, see Gunderson and Holling, *Panarchy*.

32 Holling, "Understanding the Complexity."

33 Walker and Salt, *Resilience Practice*, 13; Brian Walker, Lance Gunderson, Ann Kinzig, Carl Folke, Steve Carpenter, and Lisen Schultz, "A Handful of Heuristics and Some Propositions for Understanding Resilience in Social-Ecological Systems," *Ecology and Society*, 11, no. 1 (2006): 13.

34 Some resilience thinkers clearly recognize the possibility of and need for transformation. In "A Handful of Heuristics," for example, Walker et al. distinguish different goals and statuses of systems in adaptive cycles—transformation, resilience, adaptation—noting lack of knowledge regarding back loop processes and proposing research into transformation and reorganization. See also Brian Walker, C. S. Holling, Stephen R. Carpenter, and Ann Kinzig, "Adaptability and Transformability in Social-Ecological Systems," *Ecology and Society* 9, no. 2 (2004): 5. http://www.ecologyandsociety.org/vol9/iss2/art5/; Carl Folke, Stephen R. Carpenter, Brian Walker, Marten Scheffer, Terry Chapin, and Johan Rockström, "Resilience Thinking: Integrating Resilience, Adaptability and Transformability," *Ecology and Society* 15, no. 4 (2010): 20. http://www.ecologyandsociety.org/vol15/iss4/art20/. Drawing on this work, Per Olsson, Victor Galaz, and Wiebren J. Boonstra, "Sustainability Transformations: A Resilience Perspective," *Ecology and Society* 19, no. 4: 1. Doi: 10.5751/ES-06799-190401 survey diverse studies of transformation in resilience thinking and transition management and call for more work on the matter. While most cases of transformation studied entailed unintended regime shifts into "degraded" new regimes, Olsson, Galaz, and Boonstra call for more work on "actively navigated socio-ecological transformations" (2014, n.p.). See also Per Olsson, Carl Folke, Thomas Hahn, "Social-Ecological transformation for Ecosystem Management: The Development of Adaptive Co-Management of a Wetland Landscape in Southern Sweden," *Ecology and Society* 9, no. 2 (2004). 10.5751/ES-00683-090402.

35 Holling, "From Complex Regions," 5.

36 Paul Crutzen, "Geology of Mankind," *Nature* 415, no. 6867 (2002): 23. Doi:10.1038/415023a; Steffen, Crutzen and McNeill, "The Anthropocene," 616.

37 Steffen, Crutzen, and McNeill, "The Anthropocene," 617.

38   1610 proposal: Simon L. Lewis and Mark A. Maslin, "Defining the Anthropocene," *Nature* 519 no. 7542 (2015): 171-180. Doi: 10.1038/nature14258. Great Acceleration proposal: Anthropocene Working Group, Media note. August 29, 2016. http://www2.le.ac.uk/offices/press/press-releases/2016/august/media-note-anthropocene-working-group-awg

39   Jan Zalasiewicz, Mark Williams, Colin N. Waters, Anthony D. Barnosky, Peter Haff, "The Technofossil Record of Humans," *The Anthropocene Review*, 1, no. 1 (2014): 34–43.

40   Daniel Hartley, "Against the Anthropocene," *Salvage*, 2015. http://salvage.zone/in-print/against-the-anthropocene/.

41   Anyway, as Finney and Edwards write, speaking of the similarity between the Anthropocene and the Renaissance, "Without doubt, scholars have argued over the singular human creation, whether in literature, architecture, or art, that initiated the Renaissance, but there is no need to define its beginning, because the dates and locations of the creations are well established. Furthermore, it would be contrary to current practice to define its beginning at a single point in time because it is a cultural movement that is not tied to a single date. The same is true for the Anthropocene, whether it is a hydroelectric dam constructed in the Italian Alps, a gold mine in South Africa, the dramatic increase in carbon combustion during the Industrial Revolution, the growth of a megacity, the clearing of rain forests, or the increase in CO in the atmosphere and the resulting increase in global surface temperatures. Is putting an official beginning on the Anthropocene any more advantageous than on the Renaissance? The only reason appears to be to give it credence as a unit of the geologic time scale." Stanley Finney and Lucy Edwards, "The 'Anthropocene' Epoch: Scientific Decision or Political Statement?" *GSA Today*, 26, no. 3 (2016): 8.

42   To be clear, what I propose is not a timeline, alternate periodization, or golden spike. The adaptive cycle is rather a heuristic, a vision device that helps us see contemporary situations and practices in a different light and open other imaginaries.

43   Robert Marks, *The Origins of the Modern World: A Global and Ecological Narrative from the Fifteenth to the Twenty-First Century* (Lanham, Md: Rowman & Littlefield, 2007); Steffen, Crutzen, McNeill, "The Anthropocene," 616. Steffen, Crutzen, McNeill, "The Anthropocene," suggest we understand the Anthropocene in terms of stages, with Stage 1 being The Industrial Era (ca. 1800–1945); The Great Acceleration (1945–ca. 2015): Stage 2; and finally, their hopeful version of Stage 3, "Stewards of the Earth System? (ca. 2015–?): Stage 3 of the Anthropocene."

44 Neil Smith, *Uneven Development: Nature, Capital, and the Production of Space* (Athens, GA and London: University of Georgia Press, 1990).

45 Bruce Braun and Noel Castree, *Social Nature: Theory, Practice and Politics* (London: Wiley- Blackwell, 2001); Neil Smith, "The Production of Nature," in *Futurenatural: Nature, Science, Culture*, edited by Jon Bird, Barry Curtis, Melinda Mash, Tim Putnam, George Robertson, Lisa Tickner (London: Routledge, 1996), 35–54.

46 Matthew Gandy, *Concrete and Clay: Reworking Nature in New York City* (Cambridge, MA: MIT Press, 2003), 141.

47 Michel Foucault, *Discipline and Punish: The Birth of the Prison* (New York: Pantheon Books, 1977); Timothy Mitchell, *Colonizing Egypt* (Berkeley, CA: University of California Press, 1988).

48 Jeremy Brecher, *Strike!* (Cambridge, MA: South End Press, 1997).

49 Martin Glaberman, *Punching Out & Other Writings* (Chicago: Charles H. Kerr Publishing Company, 2002).

50 Kenneth T. Jackson, *Crabgrass Frontier: The Suburbanization of the United States* (New York: Oxford University Press, 1985).

51 On the structuring of anti-colonial movements in this Enlightenment framework see David Scott, *Conscripts of Modernity: The Tragedy of Colonial Enlightenment* (Durham, NC: Duke University Press, 2004).

52 John Law, "What's Wrong With a One World World?" *Distinktion: Journal of Social Theory* 16, no. 1 (2015), 126-139. Doi: 10.1080/1600910X.2015.1020066.

53 http://www.igbp.net/globalchange/greatacceleration.4.1b8ae20512db692f2a680001630.html

54 Ashley Carse, "Keyword: Infrastructure: How a Humble French Engineering Term Shaped the Modern World," in *Infrastructures and Social Complexity: A Companion*, edited by Penelope Harvey, Casper Bruun Jensen, Atsuro Morita (Abingdon: Routledge, 2017).

55 Jan Zalasiewicz and Mark Williams, *The Goldilocks Planet: The 4 Billion Year Story of Earth's Climate* (Oxford: Oxford University Press, 2012).

56 Brian Fagan, *The Long Summer: How Climate Changed Civilization* (London: Granta Books, 2004), 25.

57 Francis Fukuyama, *The End of History and the Last Man* (New York: The Free Press, 1992).

58 Robert Steven Nerem, B. D. Beckley, John T. Fasullo, Benjamin Hamlington, D. Masters, G. T. Mitchum, "Climate-Change–Driven

Accelerated Sea-Level Rise," *Proceedings of the National Academy of Sciences* 115, no. 9 (2018), 2022-2025. Doi: 10.1073/pnas.1717312115.

59   Ibid.

60   Steffen, Crutzen, and McNeil, "The Anthropocene."

61   Jefferson Cowie, *Stayin' Alive: The 1970s and the Last Days of the Working Class* (New York: The New Press, 2010); Loïc Wacquant, "Class, Race and Hyperincarceration in Revanchist America," *Daedalus* 139, no. 3, (2010), 74-90.

62   On surplus population concept, Aaron Benanav and John Clegg, "Misery and Debt: On the Logic and History of Surplus Populations and Surplus Capital," *Endnotes*, 2, 2010. https://endnotes.org.uk/issues/2/en/endnotes-misery-and-debt; Aaron Benanav, *A Global History of Unemployment Since 1949* (London: Verso, forthcoming).

63   Déborah Danowski and Eduardo Viveiros de Castro, *The Ends of the World* (Cambridge, UK: Polity Press, 2017); Roy Scranton, *Learning to Die in the Anthropocene: Reflections on the End of a Civilization* (San Francisco, CA: City Lights Publishers, 2015).

64   See John Robb's highly recommended Twitter (@johnrobb) and Global Guerillas Report (https://www.patreon.com/johnrobb, as well as John Robb, *Brave New War: The Next Stage of Terrorism and the End of Globalization* (Hoboken, NJ: Wiley, 2008).

65   Joshua Citarella, *Politigram and the Post-Left*, short version, 2018. http://joshuacitarella.com/_pdf/Politigram_Post-left_2018_short.pdf.

66   Thomas Meaney, "Populist Insurgency," *The New Yorker*, February 26, 2018. https://www.newyorker.com/magazine/2018/02/26/a-celebrity-philosopher-explains-the-populist-insurgency.

67   Claire Colebrook and Jami Weinstein, "Preface: Postscript on the Posthuman," in *Posthumous Life: Theorizing Beyond the Posthuman*, edited by Jami Weinstein and Claire Colebrook (New York, NY: Columbia University Press, 2017), xxiii.

68   Evans and Reid, *Resilient life*, 203.

69   Manuel Castells, *Networks of Outrage and Hope: Social Movements in the Internet Age* (Cambridge: Polity, 2015); Angela Nagle, *Kill All Normies: Online Culture Wars from 4chan and Tumbler to Trump and the Alt-Right* (Zero Books, 2017); The Invisible Committee, *Now* (South Pasadena, CA: Semiotext(e), 2017).

70   Kyle C. Cavanaugh, "Poleward Expansion of Mangroves is a Threshold Response to Decreased Frequency of Extreme Cold

Events," *Proceedings of the National Academy of Sciences*, 111, no. 2 (2014): 723—727.

71 McClatchy, "Everglades Python Challenge Wraps Up," *Miami Herald*, February 15, 2016. https://www.miamiherald.com/news/local/environment/article60450246.html.

72 Emma Marris, "How a Few Species Are Hacking Climate Change," *National Geographic*, May 6, 2014. https://news.nationalgeographic.com/news/2014/05/140506-climate-change-adaptation-evolution-coral-science-butterflies/; I-Ching Chen, Jane K. Hill, Ralf Ohlemüller, David B. Roy, Chris D. Thomas, "Rapid Range Shifts of Species Associated with High Levels of Climate Warming," *Science*, 333, no. 6045 (2011): 1024—1026.

73 Nigel Clark and Kathryn Yusoff, "Geosocial Formations and the Anthropocene." *Theory, Culture & Society* 34, no. 2-3 (2017): 3–23. doi:10.1177/0263276416688946; Alison Flood, "'Post-truth' Named Word of the Year by Oxford Dictionaries," *The Guardian*, November 15, 2016. https://www.theguardian.com/books/2016/nov/15/post-truth-named-word-of-the-year-by-oxford-dictionaries.

74 Richard Grusin, ed. *Anthropocene Feminism*. Minneapolis: University of Minnesota Press, 2017); Donna Haraway, *Staying with the Trouble: Making Kin in the Chthulucene* (Durham, NC: Duke University Press, 2016); Kathryn Yusoff, *A Billion Black Anthropocenes or None* (Minneapolis: University of Minnesota Press, 2019); Jason Moore, ed., *Anthropocene or Capitalocene? Nature, History, and the Crisis of Capitalism* (Oakland, CA: PM Press, 2015); Andreas Malm and Alf Hornborg, "A Geology of Mankind? A Critique of the Anthropocene Narrative," *The Anthropocene Review* 1, no. 1 (2014): 62–59.

75 Holling, "Resilience and Life," 2011.

76 Holling, "From Complex Regions," 5.

77 Walker and Salt, *Resilience Thinking*, 2006, xiii.

78 Holling's 1973 theory has been "tested" in numerous and diverse ecosystems, all seen as confirming its veracity. International conferences such as the Global Forum on Urban Resilience and Adaptation or the Resilience Alliance/Stockholm Resilience Centre's Resilience Summit gather scientists and practitioners from around the world to debate and share research on resilience.

79 For other versions of this argument see Stephanie Wakefield, "Inhabiting the Anthropocene Back Loop," *Resilience: International Policies, Practices and Discourses*, 6, no. 1 (2017): 1-18; Wakefield and Braun, "Oystertecture," 2019.

80 Johan Rockström, Will Steffen, Kevin Noone, Åsa Persson, F. Stuart III Chapin, Eric Lambin, Timothy M. Lenton, Marten Scheffer, Carl Folke, Hans Joachim Schellnhuber, Björn Nykvist, Cynthia A. de Wit, Terry Hughes, Sander van der Leeuw, Henning Rodhe, Sverker Sörlin, Peter K. Snyder, Robert Costanza, Uno Svedin, Malin Falkenmark, Louise Karlberg, Robert W. Corell, Victoria J. Fabry, James Hansen, Brian Walker, Diana Liverman, Katherine Richardson, Paul Crutzen, and Jonathan Foley, "Planetary boundaries: Exploring the Safe Operating Space for Humanity," *Ecology and Society*, 14, no. 2 (2009): 32. http://www.ecologyandsociety.org/vol14/iss2/art32/.

81 Johan Rockström, "Let the Environment Guide our Development," Video. TED Talk. TEDGlobal. 2010. https://www.ted.com/talks/johan_rockstrom_let_the_environment_guide_our_development.

82 Ibid. Most recently Rockström spoke at the World Economic Forum and United Nations, where he called for global institutional collaboration to manage these thresholds, framed as a paradigm shift able to combine environmental conservation and economic development. Johan Rockström, "Beyond the Anthropocene," Video file, 2017. http://www.stockholmresilience.org/research/research-news/2017-02-16-wef-2017-beyond-the-anthropocene.html.

83 Rockström et al., "Planetary Boundaries," 2.

84 Cynthia Rosenzweig and William Solecki, "Hurricane Sandy and Adaptation Pathways in New York: Lessons from a First-Responder City," *Global Environmental Change* 28, no. 1 (2014): 395-408. Doi:10.1016/j.gloenvcha.2014.05.003.

85 Judith Rodin, "Realizing the Resilience Dividend," Rockefeller Foundation, January 22, 2014. https://www.rockefellerfoundation.org/blog/realizing-resilience-dividend/.

86 The Rockefeller Foundation, "The Resilience Age," Film, 2016. https://www.youtube.com/watch?v=w-wDyhewNZ0.

87 Michel Foucault, "The Confession of the Flesh," in *Power/Knowledge: Selected Interviews and Other Writings*, edited by Colin Gordon, 194-228 (New York: Pantheon Books, 1980), 194.

88 Clive Barnett and Gary Bridge, "Thinking Problematically About the City," *International Journal of Urban and Regional Research* 40, 6 (2016): 1186–1204; Bruce Braun, "A New Urban Dispositif? Governing Life in an Age of Climate Change," *Environment and Planning D: Society and Space*, 32, no. 1 (2014): 49-64.

89 For diverse studies of liberal urban government and its subjectivity as produced differently across place and time see Christine Boyer, *Dreaming the Rational City: The Myth of American City Planning*

(Cambridge: MIT Press, 1986); Thomas Osborne, "Security and Vitality: Drains, Liberalism, and Power in the Nineteenth Century," in *Foucault and Political Reason: Liberalism, Neo-Liberalism, and Rationalities of Government* edited by Andrew Barry, Nikolas Rose, and Thomas Osborne, 99-121 (Chicago: University of Chicago Press, 1996); Chris Otter, "Making Liberalism Durable: Vision and Civility in the Late Victorian City," *Social History*, 27, no. 1 (2002): 1-15.

90 Mitchell, *Colonizing Egypt*, 1988.

91 Patrick Joyce, *The Rule of Freedom: The City and Modern Liberalism* (London: Verso, 2003).

92 Taming nature's muddy paths, Rudolf Mrázek, *Engineers of Happy Land: Technology and Nationalism in a Colony* (Princeton, NJ: Princeton University Press, 2002); on labor discipline via pipelines, see Tim Mitchell, *Carbon Democracy: Political Power in the Age of Oil* (London and New York: Verso, 2013); on labor discipline via automation, see David Noble, *Forces of Production: A Social History of Industrial Automation* (Oxford: Oxford University Press, 1984).

93 Mike Davis, *Ecology of Fear: Los Angeles and the Imagination of Disaster* (New York: Metropolitan Books, 1998).

94 Katherine Beckett and Steve Herbert, "Dealing with Disorder: Social Control in the Post-Industrial City," *Theoretical Criminology*, 12, no. 1 (2008): 5-30; Neil Smith, *The New Urban Frontier: Gentrification and the Revanchist City* (London and New York: Routledge, 1996).

95 On the transformation of New York in this period see Miriam Greenberg, *Branding New York: How a City in Crisis was Sold to the World* (New York: Routledge, 2008); Peter Eisinger, "The Politics of Bread and Circuses: Building the City for the Visitor Class," *Urban Affairs Review* 35, no. 3 (2000): 316-333. Doi: 10.1177/107808740003500302.

96 Mike Davis, *City of Quartz: Excavating the Future in Los Angeles* (London and New York: Verso, 1990), 223.

97 Stephen Graham, *Cities under Siege: The New Military Urbanism* (New York and London: Verso, 2011).

98 For discourse of brittle infrastructure vulnerability, see Amory B. Lovins and L. Hunter Lovins, *Brittle Power: Energy Strategy for National Security* (Denver: Brick House Pub Company, 1982). On emergence and cascading infrastructural risk, Myriam Dunn Cavelty and Kristian Soby Kristensen, *Securing "the Homeland": Critical Infrastructure, Risk and (In)Security* (London and New York: Routledge, 2008); Michael Dillon, "Governing Terror: The State of Emergency of Biopolitical Emergence," *International Political Sociology*, 1 no. 1 (2007): 7-28; Brian Massumi, "National Enterprise

Emergency: Steps Toward an Ecology of Powers," *Theory, Culture & Society* 26, no. 6 (2009): 153-185.

99 Jon Coaffee, "Risk, Resilience and Environmentally Sustainable Cities," *Energy Policy*, 36, no. 12, (2008), 4633-4638: 4633.

100 Stephen Graham, "Urban Metabolism as Target: Contemporary War as Forced Demodernization," in *The Nature of Cities: Urban Political Ecology and the Politics of Urban Metabolism*, edited by Nik Heynen, Maria Kaika, and Erik Swyngedouw, 234-252 (London: Routledge, 2006).

101 Ibid, 261.

102 City of New York, "Mayor Bloomberg, police commissioner Kelly and Microsoft unveil new, state-of-the-art law enforcement technology that aggregates and analyzes existing public safety data in real time to provide a comprehensive view of potential threats and criminal activity," (Press release, New York, August 8). August 8, 2012. http://www.nyc.gov/ portal/site/nycgov/menuitem.c0935b9a57bb4ef3daf2f1c701c789a0/index.jsp? pageID=mayor_press_release&catID=1194&doc_name=http://www.nyc.gov/html/om/ html/2012b/pr291-12.html&cc=unused1978&rc=1194&ndi=1; Olivia J. Greer, "No Cause of Action: Video Surveillance in New York City," *Michigan Telecommunications and Technology Law Review*, 18, no. 2 (2012): 589-626.

103 Michael Bloomberg, "Mayor Bloomberg delivers address on shaping New York City's future after Hurricane Sandy," Press conference, New York, Marriot Downtown, December 6, 2012. http://www.nyc.gov/portal/site/nycgov/ menuitem.c0935b9a57bb4ef3daf2f1c701c789a0/index.jsp? pageID=mayor_press_release&catID=1194&doc_name=http://www.nyc.gov/html/om/ html/2012b/pr459-12.html&cc=unused1978&rc=1194&ndi=1; Richard Florida and Sara Johnson, "Making Our Coastal Cities More Resilient Can't Wait," *The Atlantic City Lab*, November 1, 2012. http://www.citylab.com/work/2012/11/making-our-cities-more-resilient-cant-wait/3758/.

104 Baden Copeland, Josh Keller, and Bill Marsh, "What Could Disappear," *New York Times*, November 24, 2012. http://www.nytimes.com/interactive/2012/11/24/opinion/sunday/ what-could-disappear.html

105 Tim Folger, "Rising Seas," *National Geographic*, September, 2013. http://ngm.nationalgeographic.com/2013/09/rising-seas/folger-text; New York City Panel on Climate Change, "Climate Risk Information 2013: Observations, Climate Change Projections, and Maps." Report. 2013. http://www.nyc.gov/html/planyc2030/downloads/pdf/npcc_climate_risk_information_2013_report.pdf; Jeff Tollefson, "Natural Hazards: New York vs. the

Sea," *Nature*, February 13, 2013. http://www.nature.com/news/natural-hazards-new-york-vs-the-sea-1.12419.

106 Bloomberg, "Mayor Bloomberg Delivers."

107 Andrew Zolli, "Learning to Bounce Back," *New York Times*, November 2, 2012. http://www.nytimes.com/2012/11/03/opinion/forget-sustainability-its-about-resilience.html.

108 Many observers noted the inability of "sustainable" infrastructures to stop Sandy's destruction, highlighting for example how, despite having the largest number of LEED certified buildings in the world, the financial district was still underwater and in the dark. Richard Florida and Andrew Zolli, "The Rush to Resilience: We Don't Have Decades Before the Next Sandy," *The Atlantic City Lab*, November 9, 2012. http://www.citylab.com/work/2012/11/building-resilient-cities-conversation-andrew-zolli-and-jonathan-rose/3839/.

109 Zolli, "Learning to Bounce."

110 Michael Bloomberg, "Mayor Bloomberg Presents the City's Long-term Plan to Further Prepare for The Impacts of a Changing Climate" (Press conference, New York, June 11, 2013). http://www1.nyc.gov/office-of-the-mayor/news/200-13/mayor-bloomberg-presents-city-s-long-term-plan-further-prepare-the-impacts-a-changing.

111 Adams, "Notes from the Resilient City," *Log*, 32 (2014).

112 Zolli, "Learning to Bounce." The headline, still present in the article URL, has since been changed to "Learning to bounce back." See https://www.nytimes.com/2012/11/03/opinion/forget-sustainability-its-about-resilience.html.

113 On resilience in the words of its proponents see Carl Folke, "Resilience: The Emergence of a Perspective For Social–Ecological Systems Analyses," *Global Environmental Change* 16 (2006): 253–267; Holling, 1973. On urban resilience and its systems thinking perspective see Jon Coaffee and Peter Lee, *Urban Resilience: Planning for Risk, Crisis, and Uncertainty* (New York and London: Palgrave, 2017); Sara A. Meerow, Joshua P. Newell, and Melissa Stults, "Defining Urban Resilience: A Review," *Landscape and Urban Planning* 147 (2016): 38–49.

114 There is certainly a parallel here with what Naomi Klein has called the capitalist "shock doctrine." Naomi Klein, *The Shock Doctrine: The Rise of Disaster Capitalism* (Toronto: A.A. Knopf Canada, 2007). The drive to "Fix&Fortify" New York, as the city's campaign is called, was equally an investment opportunity, and "the rush to resilience" brought the city's wealthiest forces out en masse, each seeking a piece of the opportunity posed by crisis as "the new normal."

115 Cynthia Rosenzweig and William Solecki, "Building Climate Resilience in Cities: Lessons from New York," *The Conversation*, January 22, 2016. https://theconversation.com/building-climate-resilience-in-cities-lessons-from-new-york-52363.

116 On New York's long history of being a site for experimentation with forms of design, government, and thought that would later be exported globally, see Rem Koolhaas, *Delirious New York: A Retroactive Manifesto for Manhattan* (New York: Monacelli Press, 1997).

117 Michael Berkowitz, "The Movement We're Building," *100 Resilient Cities*, July 24, 2017). https://www.100resilientcities.org/the-movement-were-building/. On 100 Resilient Cities, see https://www.100resilientcities.org/about-us/#section-2.

118 Jeff Goodell, "Miami: How Rising Sea Levels Endanger South Florida," *Rolling Stone*, June 20, 2013. https://www.rollingstone.com/politics/politics-news/miami-how-rising-sea-levels-endanger-south-florida-200956/.

119 Rockefeller Foundation, "100 Resilient Cities Announces Global Summit—Largest Ever Gathering of Urban Resilience Experts" (Press release, June 23, 2017).

120 Holling, "From Complex Regions," 7.

121 Ibid, 7.

122 On urban living laboratories, see Harriet Bulkeley, Simon Marvin, Yuliya Voytenko Palgan, Kes McCormick, Marija Breitfuss-Loidl, Lindsay Mai, Timo von Wirth, and Niki Frantzeskaki, "Urban Living Laboratories: Conducting the Experimental City?" *European Urban and Regional Studies* (2018). Doi:10.1177/0969776418787222; James Evans and Andrew Karvonen, "Give Me a Laboratory and I Will Lower Your Carbon Footprint!'—Urban Laboratories and the Governance of Low-Carbon Futures," *International Journal of Urban and Regional Research* 38, no. 2 (2014): 413-430.

123 Rebuild by Design, "The Rebuilders." Video file. 2014. https://vimeo.com/90825595.

124 Rebuild by Design, "Promoting Resilience Post-Sandy Through Innovative Planning and Design," Rebuild by Design: Hurricane Sandy Regional Planning and Design Competition. Design brief. June 21, 2013, 1. http://portal.hud.gov/hudportal/documents/huddoc?id=REBUILDBYDESIGNBrief.pdf

125 City of New York, "PlaNYC: A Stronger, More Resilient New York." Report prepared by the Special Initiative for Rebuilding and Resiliency. 2013, 5-6. http://www.nyc.gov/html/sirr/html/report/report.shtml. Federal Emergency Management Association (FEMA),

"A Whole Community Approach to Emergency Management: Principles, Themes, and Pathways for Action," Washington, DC: US Department of Homeland Security, 2011. https://www.fema.gov/media-library-data/20130726-1813-25045-0649/whole_community_dec2011__2_.pdf.

126 Andrew Cuomo, "We Will Lead on Climate Change," *New York Daily News,* November 25, 2012. http://www.nydailynews.com/opinion/lead-climate-change-article-1.1202221

127 City of New York, "PlaNYC," 157.

128 City of New York, "PlaNYC: A Stronger," i.

129 New York City Office of Emergency Management, "Meet Ready Girl." 2019. https://www1.nyc.gov/site/em/ready/ready-girl.page.

130 City of New York "PlaNYC," 157.

131 Michael Kane, "Why New York Should Become the City of Oysters Again." *New York Post,* June 21, 2014. https://nypost.com/2014/06/21/why-new-york-should-become-the-city-of-oysters-again/.

132 SeArc - Ecological Marine Consulting, "About SeArc." 2010. http://www.searc-consulting.com/home.yecms/index; Shimrit Perkol-Finkel and Ido Sella, "Ecologically Active Concrete for Coastal and Marine Infrastructure: Innovative Matrices and Designs." Proceeding of the 10th ICE Conference: from sea to shore - meeting the challenges of the sea. 2014. Doi: 10.1680/fsts597571139

133 SCAPE / Landscape Architecture, *Rebuild by Design/ Living Breakwaters (IP Report, Staten Island and Raritan Bay)* (New York: Rebuild by Design, 2013): 23.

134 Nicholas Nehamas, "Miami's Downtown Building Boom Drawing to a Close," Miami Herald. October 14, 2015. http://www.miamiherald.com/news/business/real-estate-news/article39189630.html; Alejandro Portes and Ariel C. Armony, *The Global Edge: Miami in the 21st Century* (Oakland, CA: University of California Press, 2018); Anthony Adragna, "Florida Senator Nelson Calls Sea-Level Rise, Climate Change "Compelling Story," Announces Hearing," *Bloomberg BNA,* March 12, 2014. https://www.bna.com/florida-sen-nelson-b17179882772/.

135 Erika Spanger-Siegfried, Melanie Fitzpatrick, and Kristina Dahl, "Encroaching Tides: How Sea Level Rise and Tidal Flooding Threaten US East and Gulf Coast Communities Over the Next 30 Years." (Cambridge, MA: Union of Concerned Scientists, 2014). http://www.ucsusa.org/encroachingtides; Southeast Florida Regional Climate Change Compact (SFRCCC), "Unified Sea Level Rise Projection for Southeast Florida," October 2015.

http://www.southeastfloridaclimatecompact.org/ wp-content/uploads/2015/10/2015-Compact-Unified-Sea-LevelRise-Projection.pdf; Shimon Wdowinski, Ronald Bray, Ben P. Kirtman, and Zhaohua Wu, "Increasing Flooding Hazard in Coastal Communities Due to Rising Sea Level: Case study of Miami Beach, Florida." *Ocean & Coastal Management*, 126 (2016): 1-8. Doi: 10.1016/j.ocecoaman.2016.03.002.

136 SFRCCC, 2015; Spanger-Siegfried, Fitzpatrick, and Dahl, "Encroaching Tides." https://www.ucsusa.org/sites/default/files/attach/2016/04/miami-dade-sea-level-rise-tidal-flooding-fact-sheet.pdf.

137 Tristram Korten, "In Florida, Officials Ban Term 'Climate Change'," *Miami Herald*, March 8, 2015. https://www.miamiherald.com/news/state/florida/article12983720.html.

138 Kate Stein, "'We're a Living Laboratory': Miami Beach Works on Resiliency as Businesses Face Flooding," WLRN.org, October 17, 2016. http://www.wlrn.org/post/were-living-laboratory-miami-beach-works-resiliency-businesses-face-flooding

139 Greater Miami and the Beaches, "Preliminary resilience assessment." Report created for 100 Resilient Cities. 2017. http://www.mbrisingabove.com/wp-content/uploads/2017/10/170905_GMB-PRA_v01.pdf; Thaddeus Pawlowski and Michelle Mueller, "The Resilience Accelerator—Getting up to Speed in Southeast Florida." 100 Resilient Cities blog. November 20, 2018. https://www.100resilientcities.org/resilience-accelerator-getting-up-to-speed-southeast-florida/. Urban Land Institute (ULI), "Stormwater Management and Climate Adaptation Review." An Urban Land Institute Advisory Services Panel Report, Miami Beach, FL. April 16-19, 2018. http://www.mbrisingabove.com/wp-content/uploads/2018/04/Miami-Beach_Panel_Report_lo-res.pdf. In a report designed to help the City improve on the negative public relations generated by its resiliency projects, ULI has recommended Miami Beach brand itself as the "Resilient Art City"—by working with artists and local cultural institutions to paint pumps or design aesthetics on their infrastructure and creating "iconic pilot projects" (ULI, 2018, 49).

140 ULI, 2018; Philip Levine. Philip Levine for Mayor of Miami Beach "Paddle." Campaign ad [video]. 2013. https://www.youtube.com/watch?v=N9niAnh9KZw.

141 See City of Miami Beach Capital Improvement Projects site for official description of projects by area: https://www.miamibeachfl.gov/city-hall/cip/.

142 Joey Flechas and Jenny Staletovich, "Miami's Battle to Stem Rising Tides." *Miami Herald*, October 23, 2015. https://www.miamiherald.com/news/local/community/miami-dade/miami-beach/article41141856.html; Alex Harris, "Miami Beach Wants Higher Roads and Pumps to Fight Sea Rise. Some Residents Say No Way." *Miami Herald*, May 16, 2018. https://www.miamiherald.com/news/local/community/miami-dade/miami-beach/article211237324.html

143 Todd A. Crowl and Rita A. Teutonico, ""As the Sea Rises, South Floridians Will Get Thirsty Before They Get Wet," *Sun Sentinel*, June 4, 2018. https://www.sun-sentinel.com/opinion/commentary/fl-op-viewpoint-sea-level-rise-freshwater-supply-20180601-story.html.

144 https://evergladesrestoration.gov/.

145 Deering Estate, "Rehydration Project," 2019. https://deeringestate.org/conservation/rehydration-project/.

146 Amanda Ruggeri, "Miami's Fight Against Rising Seas." *BBC Future Now*. April 4. 2017. http://www.bbc.com/future/story/20170403-miamis-fight-against-sea-level-rise. Much of this new construction is aimed at wealthy foreign cash buyers from Latin America, Russian oligarchs, as well as "creative class" young professionals.

147 Thomás Regalado, "We cannot allow this to become the new 'normal'." *Twitter*. August 2, 2017.

148 Walker and Salt, 2012, 13.

149 Fath, Dean, and Katzmair, "Navigating the Adaptive," 3.

150 Holling, "From Complex Regions," 7; 1.

151 Fath, Dean, and Katzmair, "Navigating the Adaptive."

152 Gunderson, Holling, and Light, *Barriers and Bridges*.

153 Sandi Zellmer and Lance Gunderson, "Why Resilience May Not Always Be a Good Thing: Lessons in Ecosystem Restoration from Glen Canyon and the Everglades," *Nebraska Law Review* 87, no. 4 (2009): 893. https://ssrn.com/abstract=1434386.

154 Ariane Lourie Harrison, *Architectural Theories of the Environment: Posthuman Territory* (New York: Routledge, 2013).

155 Holling, "Resilience and Stability"; Gunderson and Holling, *Panarchy;* Folke, "Resilience."

156 Folke, "Resilience," 253; 256.

157 Fikret Berkes and Carl Folke, eds., *Linking Social and Ecological Systems: Management Practices and Social Mechanisms for Building Resilience* (Cambridge University Press, New York, 1998).

158 Folke, "Resilience," 255.

159 Fath, Dean, and Katzmair, "Navigating the Adaptive," 1. See also the work of the Stockholm Resilience Centre, the United Nations, Rebuild by Design, and 100 Resilient Cities.

160 Gunderson, Interview with Stephanie Wakefield, 2019.

161 This section builds on arguments advanced in Wakefield and Braun, "Oystertecture."

162 Tim McDonnell, "Meet the Woman Tasked with Saving New York From the Next Sandy," *CityLab*, December 10, 2012. https://www.citylab.com/equity/2012/12/can-woman-save-new-york-next-sandy/4115/.

163 Stephanie Wakefield and Bruce Braun, "Oystertecture," 2019.

164 Stephanie Wakefield, "Miami Beach Forever? Urbanism in the Back Loop," *Geoforum*, 107 (2019): 33-44.

165 This view of the resilience industry as a cynical attempt to capture money and maintain the status quo is in fact now shared by some of resilience ecology's founding thinkers. Gunderson, Interview with Stephanie Wakefield, 2019.

166 Kate Orff, Interview with Stephanie Wakefield, SCAPE Office, New York, NY, May 29, 2015.

167 Australian Government, "Highlights of the Reef 2050 Long-Term Sustainability Plan." 2015. http://www.environment.gov.au/marine/gbr/publications/highlights-long-term-sustainability-plan#.

168 Adams, "Notes."

169 BIG Bjarke Ingels Group, "The Big U." Final proposal submitted to Rebuild by Design. 2014. http://www.rebuildbydesign.org/data/files/675.pdf.

170 Ross Exo Adams, "Becoming-Infrastructural," *e-flux architecture*, 2017. http://www.e-flux.com/architecture/positions/149606/becoming-infrastructural/; Evans and Reid, *Resilient Life*.

171 Kevin Grove, "From Emergency Management to Managing Emergence: A Genealogy of Disaster Management in Jamaica," *Annals of the Association of American Geographers*, 103, no. 3 (2013): 583.

172 Orff, Interview.

173 One might note that urban resilience ambitions seem to be growing. In addition to opportunities for investment and recalibration of governance, urban resilience is increasingly seen as stepping in

Notes

to replace what is seen as failed national governments. As 100RC President Michael Berkowicz proclaims, "This work is critical now, as cities are stepping forward to fill the leadership void left by national governments, and it will only grow in importance as the world gets drastically more urbanized over the coming decades." Worth noting, as cities along with corporations and other non-state territorial powers become key contenders for power in coming decades.

174 Steffen, et al., "Trajectories of the Earth," 2018.

175 Carl Schmitt called the *katechon* the permanent management of the present to hold back the forces of chaos, a role he saw played by different political authorities at different times: the Holy Roman Empire, the Byzantine Empire, or individual authorities such as Emperor Rudolf II of Hapsburg. Carl Schmitt, *The Nomos of the Earth in the International Law of the Jus Publicum Europaeum* (New York: Telos, 2003). Giorgio Agamben expands the range even further. For him, "every theory of the State, including Hobbes's—which thinks of it as a power destined to block or delay catastrophe—can be taken as a secularization of this interpretation of 2 Thessalonians 2." Giorgio Agamben, *The Time That Remains* (Stanford: Stanford University Press, 2005), 110. As I understand it, katechonic is any force that depicts reality in terms of a false binary of catastrophe and chaos, on one side, and that force's own imposition of order, on the other. In this blackmail situation—which historically was imposed by the Church vis a vis the end times but is also that of current political regimes regarding climate change or revolution—existing regimes lock populations into a hostage situation by portraying themselves as the only legitimate bulwark against certain catastrophe. For further discussion of the transformation resilience poses for the *katechon* see Wakefield and Braun, "Oystertecture;" See also Stephanie Wakefield and Bruce Braun, "Inhabiting the Post-Apocalyptic City," *Society and Space Open*, 2014. https://societyandspace.org/2014/02/11/inhabiting-the-postapocalytic-city-bruce-braun-and-stephanie-wakefield/.

176 Marshall Sahlins, *The Western Illusion of Human Nature* (Chicago, Prickly Paradigm Press, 2008). For similar arguments regarding politics and revolution, see The Invisible Committee, *To Our Friends* (Los Angeles, CA: Semiotext(e): 2014).

177 Sahlins, *The Western Illusion*, 83.

178 Reiner Schürmann, "'What Must I Do?' at the End of Metaphysics: Ethical Norms and the Hypothesis of a Historical Closure," in *Phenomenology in a Pluralistic Context*, edited by William I. McBride and Calvin O. Schrag (Binghamton: SUNY Press, 1984): 49-64.

179 Schürmann, "'What Must I Do?'", 51.

180 Giorgio Agamben, "What is a Destituent Power?" *Environment and Planning D: Society and Space*, 32, no. 1 (2014): 65–74.

181 Gilles Deleuze with Claire Parnet, *Gilles Deleuze From A to Z*, "Joy," translated by Charles J. Stivale, DVD (New York: Semiotext(e), 2011).

182 Reiner Schürmann, *Heidegger on Being and Acting: From Principles to Anarchy* (Bloomington: Indiana University Press, 1987). Schürmann follows Friedrich Nietzsche's diagnosis: Friedrich Nietzsche, *Twilight of the Idols, or, How to Philosophize with a Hammer* (New York: Oxford University Press, 1998).

183 Bruno Latour, "Facing Gaia: Six Lectures on the Political Theology of Nature," The Gifford Lectures on Natural Religion Edinburgh, February 18-28, 2013a). http://www.bruno-latour.fr/sites/default/files/downloads/GIFFORD-SIX-LECTURES_1.pdf; Bruno Latour, "Telling friends from foes in the time of the Anthropocene," in edited by Clive Hamilton, Christophe Bonneuil & François Gemenne (editors). The Anthropocene and the global environment crisis—Rethinking modernity in a new epoch (London, Routledge, 2013b): 145-155; Bruno Latour, *Facing Gaia: Eight lectures on the new climatic regime* (Cambridge, UK; Medford, MA: Polity Press, 2017).

184 Latour, "Facing Gaia: Six Lectures," 121.

185 Ibid, 116. For other thinkers who have recently taken up Latour's use of the term "earthbound," see: Déborah Danowski and Eduardo Viveiros de Castro, *The Ends of the World* (Cambridge, UK: Polity, 2016); Donna Haraway, "Tentacular thinking: Anthropocene, Capitalocene, Chthulucene," *e-flux*, 75, September 2016, https://www.e-flux.com/journal/75/67125/tentacular-thinking-anthropocene-capitalocene-chthulucene/

186 Latour, "Facing Gaia: Six Lectures," 121.

187 Bernd Reiter, *Constructing the pluriverse: the geopolitics of knowledge* (Durham, NC: Duke University Press, 2018); Arturo Escobar, *Designs for the pluriverse: radical interdependence, autonomy, and the making of worlds* (Durham, NC: Duke University Press, 2018); Marisol De la Cadena and Mario Blaser, *A world of many worlds* (Durham, NC: Duke University Press, 2018).

188 Scranton, *Learning How to Die*.

189 Latour, "Facing Gaia: Six Lectures," 116.

190 Nietzsche, for example, saw the moment when the search for a perfect steady ground would come to an end as the "high point of

humanity," to be celebrated in a "gay science." Schürmann, "What Must I Do'?", 51.

191 Latour, *Facing Gaia: Eight Lectures*, 290; 219.

192 Latour, "Facing Gaia: Six Lectures," 103.

193 Anna Tsing, *The Mushroom at the End of the World: On the Possibility of Life in Capitalist Ruins* (Princeton, NJ: Princeton University Press, 2016): vii.

194 Tsing, *The Mushroom*, vii.

195 Clive Hamilton, "Human Destiny in the Anthropocene," in *The Anthropocene and the Global Environmental Crisis: Rethinking modernity in a new epoch*, edited by Clive Hamilton, François Gemenne, and Christophe Bonneuil (Abingdon, OX: Routledge, 2015): 32-43.

196 Bruno Latour, Isabelle Stengers, Anna Tsing, and Nils Bubandt, "Anthropologists are Talking—About Capitalism, Ecology, and Apocalypse," *Ethnos*, 83, no. 3 (2018): 587-606, Doi: 10.1080/00141844.2018.1457703.

197 Latour, "Facing Gaia: Six Lectures," 111.

198 Ibid, 136.

199 Ibid, 134.

200 Ibid, 136. Latour: "Even though so far there is no cult, not even a civic one for such a self-imposed tracing of 'planetary boundaries,' it is fascinating to imagine through what sort of public ceremonies such self-imposed limits would be sworn and enforced. The rituals to be imagined might not fill the churches, but they will shake the scientific disciplines quite a lot and extract from ethnography a rich lore of practices." "Facing Gaia: Six Lectures," 136.

201 Ibid, 111.

202 Ibid, 112, Latour's emphasis. "Extraordinary sentence," continues Latour on the same page, "as if the *anthropos* of the Anthropocene had to go through a *Bildungsroman* just like the bourgeois of the 19th century confronted with the time of revolutions."

203 Ibid, 117.

204 Latour, *Facing Gaia: Eight Lectures*, 223.

205 Ibid, 251.

206 Latour, "Facing Gaia: Six Lectures,", 119. Reminiscent of Uexkull's own discussion of ticks, Latour asserts "whereas the Atlas of the scientific revolution could hold the globe in his hand, scientists of the

Gaian counter-revolution, I am sorry to say, look more like ticks on the mane of a roaring beast" (2013a, 134).

207 Latour, *Facing Gaia: Eight Lectures*, 251.

208 Despite repeated references of agreement with the German philosopher, Latour's notion of entangled territories should be read in stark contrast to Peter Sloterdijk's philosophy of "spheres," which also speaks of *Umwelt*s and "life support systems," but does so in a way that emphasizes the diversity of life worlds, freedom to build them, and freedom found *in* them—rather than as a matter of being hostage to their sovereign control. See e.g. Peter Sloterdijk, *Bubbles: Spheres volume 1: Microsphereology* (Los Angeles, CA: Semiotext(e), 2011).

209 Latour, "Facing Gaia: Six Lectures," 132.

210 Ibid, 132.

211 Ibid, 117-118. Filmmaker Béla Tarr on *The Turin Horse*: "We just wanted to see how difficult and terrible it is when every day you have to go to the well and bring the water, in summer, in winter... All the time. The daily repetition of the same routine makes it possible to show that something is wrong with their world. It's very simple and pure." Béla Tarr, Interview with Vladan Petkovic, *cineuropa*, April 3, 2011. https://cineuropa.org/en/interview/198131/ Tarr has also described *The Turin Horse* as the last step in a development throughout his career: "In my first film I started from my social sensibility and I just wanted to change the world. Then I had to understand that problems are more complicated. Now I can just say it's quite heavy and I don't know what is coming, but I can see something that is very close—the end."

212 Tsing, *The Mushroom*, 22.

213 Ibid, 3.

214 Donna Haraway, *Staying with the Trouble: Making Kin in the Chthulucene* (Durham, NC: Duke University Press, 2016); Tsing, 2016.

215 Walter Benjamin, *Illuminations* (New York: Shocken Books, 1968): 257.

216 Eileen Crist, "On the Poverty of Our Nomenclature," *Environmental Humanities*, 3, no. 1 (2013): 131. Doi: 10.1215/22011919-3611266

217 Tsing, *The Mushroom*, vii.

218 Ibid, 20.

219 Ibid, 8.

220 Ibid, 2.

221 Ibid, 24.

222 Ibid, 2.

223 Ibid, 2.

224 Steven J. Jackson, "Rethinking Repair," in Tarleton Gillespie, Pablo J. Boczkowski, and Kristen A. Foot, *Media Technologies: Essays on Communication, Materiality and Society* Cambridge: MIT Press, 2014): 221-239.

225 Lauren Berlant, "The Commons: Infrastructures for Troubling Times," *Environment and Planning D: Society and Space* 34, no. 3 (2016): 393-419. Doi: 10.1177/0263775816645989.

226 Nicholas Beuret and Gareth Brown, "The Walking Dead: The Anthropocene as a Ruined Earth," *Science as Culture*, 26, no. 3 (2017): 330–354. Doi: 10.1080/09505431.2016.1257600; N.K. Jemisin, *The Fifth Season (The Broken Earth)* (London: Orbit Books, 2015).

227 Latour, *Facing Gaia: Eight Lectures*, 233.

228 Tsing, *The Mushroom*, 21.

229 Elizabeth A. Povinelli, *Geontologies: A Requiem to Late Liberalism* (Durham, N.C.: Duke University Press, 2016).

230 Beuret and Brown, "The Walking Dead."

231 Tsing, *The Mushroom*.

232 Berlant, "The Commons," 398.

233 Kevin Grove and David Chandler, "Introduction: Resilience and the Anthropocene: The Stakes of 'Renaturalising' Politics," *Resilience: International Policies, Practices and Discourses* 5, no. 2 (2016): 7. Doi: 10.1080/21693293.2016.1241476.

234 Jeff VanderMeer, *Borne* (Toronto, Ontario, Canada: HarperCollins Publishers Ltd, 2017).

235 VanderMeer, *Borne*, 296.

236 Ibid. 260.

237 David Chandler and Julian Reid, *The Neoliberal Subject: Resilience, Adaptation and Vulnerability* (Rowman & Littlefield International, London, 2016).

238 Friedrich Nietzsche, *The Will to Power* (New York: Random House, 1967), 7.

239 This parallels the critique Martin Heidegger made of Niezsche's attempt to overcome nihilism and metaphysics. Martin Heidegger,

"Nietzsche's Word: 'God is Dead'," in *Off the Beaten Track* (Cambridge, MA: Cambridge University Press, 2002). As Heidegger saw it, despite aims of going beyond nihilism and metaphysics, Nietzsche's efforts to do so via a thinking of being as becoming and man as overman remained trapped within metaphysics and nihilism itself. This was because, in Heidegger's view, rather than getting rid of the determination of being, Nietzsche still defined and valued it albeit by reversing it: what was God and the supersensory became will to power and the sensory, Being became becoming, etc. In this way, Heidegger argued, Nietzsche's attempt at transvaluation represented not an overcoming, but the completion or perfection, of metaphysics and nihilism.

240 Lyrics from Kendrick Lamar, "HUMBLE." *Damn* (Top Dawg Entertainment, Aftermath Entertainment and Interscope Records, 2018).

241 Jennifer Kavanagh and Michael D. Rich, "An Initial Exploration of the Diminishing Role of Facts and Analysis in American Public Life" (Santa Monica, CA: RAND Corporation, 2018). https://www.rand.org/pubs/research_reports/RR2314.html.

242 Kurt Anderson, "How America Lost Its Mind," *The Atlantic*, September 2017. https://www.theatlantic.com/magazine/archive/2017/09/how-america-lost-its-mind/534231/.

243 Gilles Deleuze with Claire Parnet, *Gilles Deleuze From A to Z*, "Joy," translated by Charles J. Stivale, DVD (New York: Semiotext(e), 2011).

244 VanderMeer, *Borne*, 317.

245 Ibid, 323.

246 Michael McCastle, "Imagination: The Mind's Contribution to Peak Performance," *Breaking Muscle*. Accessed March 19, 2019. https://breakingmuscle.com/fitness/imagination-the-minds-contribution-to-peak-performance.

247 Schürmann, "'What Must I Do?'", 51.

248 Michel Foucault, *The History of Sexuality, Volume 3: The Care of the Self* (New York: Vintage, 1988), 5.

249 C.S. Holling, "Resilience Dynamics." *Stockholm Resilience Centre TV*. Video file. November 5, 2008. https://www.youtube.com/watch?v=FhfmaXZPKEY&t=7s

250 Holling, "From Complex Regions," 4-5.

251 Holling, Ibid, 4-5.

252 Peter Kingsley, "The Elders." Audio File. http://peterkingsley.org/product/the-elders/.

253 Lance Gunderson, "Living with Uncertainty and Surprise." *Stockholm Resilience Centre TV.* Video file. January 23, 2009. https://www.youtube.com/watch?v=kqkfHjX9IsY.

254 Holling, "Resilience Dynamics."

255 "How to Survive a Disaster in a Big City - According to an Urban Prepper." NBC News. Video File. Accessed March 19, 2019. https://www.nbcnews.com/feature/101/video/how-to-survive-a-disaster-in-a-big-city-according-to-an-urban-prepper-1019859011805.

256 Neil Strauss, "Brock Pierce: The hippie king of cryptocurrency," *Rolling Stone*, July 26, 2018. https://www.rollingstone.com/culture/culture-features/brock-pierce-hippie-king-of-cryptocurrency-700213/.

257 Naomi Klein, *The Shock Doctrine*; Monte Reel, "How to Rebuild Puerto Rico," *Bloomberg Businessweek*, December 27, 2017. https://www.bloomberg.com/news/features/2017-12-14/how-to-rebuild-puerto-rico. Tactivate is one such interesting example of an entrepreneurial-military Special Operations disaster relief organization that emerged in the wake of Sandy and Maria: https://www.tactivate.com/.

258 Evan Osnos, "Doomsday Prep for the Super-Rich," *The New Yorker*, January 30, 2017. https://www.newyorker.com/magazine/2017/01/30/doomsday-prep-for-the-super-rich.

259 Jessica Bennett, "Rise of the Preppers: America's New Survivalists" *Newsweek*, December 29, 2009. https://www.newsweek.com/rise-preppers-americas-new-survivalists-75537.

260 Casey Ryan Kelly, "The Man-pocalpyse: Doomsday Preppers and the Rituals of Apocalyptic Manhood," *Text and Performance Quarterly*, 36, no. 2-3, (2016): 95-114. Doi: 10.1080/10462937.2016.1158415. For an ethnographic study of survivalism, often seen as a predecessor to prepping, see Richard G. Mitchell, *Dancing at Armageddon: survivalism and chaos in modern times.* Chicago: University of Chicago Press, 2002.

261 The Economist, "Preparing for the Apocalypse: I Will Survive," *The Economist*, December 17, 2014. https://www.economist.com/christmas-specials/2014/12/17/i-will-survive

262 Stefan Kamph, "Fallout Fanatics," *Miami New Times*, April 26, 2012. http://digitalissue.miaminewtimes.com/article/Fallout+Fanatics/1043822/108976/article.html

263 Holly Hartman, "I downloaded an app. And suddenly, was part of the Cajun Navy," *Houston Chronicle*, February 5, 2019. https://www.chron.com/local/gray-matters/article/I-downloaded-an-And-suddenly-I-was-talking-12172506.php#item-85307-tbla-15; "Louisiana Cajun Navy." Facebook profile. Accessed March 22, 2019. https://www.facebook.com/LaCajunNavy.

264 Michael Dillon and Julian Reid, *The Liberal Way of War: Killing to Make Life Live* (London: Routledge, 2009); Stephanie Wakefield, "Infrastructures of Liberal Life: From Modernity and Progress to Resilience and Ruins," *Geography Compass*, 12, no. 7 (2018). Doi: 10.1111/gec3.12377; Shannon Mattern, "Infrastructural Intelligence." Presentation, 5th International LafargeHolcim Forum, April 7–9, 2016. http://www.wordsinspace.net/shannon/2016/01/01/infrastructural-intelligence/; Easterling, *Extrastatecraft: The Power of Infrastructure Space* (New York & London: Verso, 2014); Jane Schneider and Ida Susser, *Wounded cities: destruction and reconstruction in a globalized world* (Oxford; New York: Berg, 2003).

265 Emily Eakin, "The Civilization Kit," *The New Yorker*, December 23 and 30, 2013. https://www.newyorker.com/magazine/2013/12/23/the-civilization-kit; "Machines: Global Village construction set." Open Source Ecology. Accessed March 19, 2019. https://www.opensourceecology.org/gvcs/.

266 Marcin Jakubowski, "Hacking the Farm with Low-Cost, Open Source Tool Designs," Interview with Don Watkins, Opensource.com, February 15, 2016. https://opensource.com/life/16/2/interview-marcin-jakubowski-open-source-ecology; YouTube channel, explore at your own risk: https://wiki.opensourceecology.org/wiki/Distillations#Factor_e_Distillations_Episode_1_-_Introduction.

267 "Marcin Jakubowski." Open Source Ecology. Accessed March 19, 2019. https://www.opensourceecology.org/marcin-jakubowski/

268 Mario Furloni, "The (Open-Source) Ecologist." Video file. Accessed March 19, 2019. https://vimeo.com/26613005

269 Eakin, "The Civilization."

270 Furloni, "The (Open-Source)."

271 Buddy Blalock, Interview with Stephanie Wakefield, Old River Landing, Louisiana, April 14, 2018.

272 Coral Davenport and Campbell Robertson, "Resettling the first American 'climate refugees.' *New York Times*, May 3, 2016. https://www.nytimes.com/2016/05/03/us/resetting-the-first-american-climate-refugees.html.

273 Jacques Lacour, Interview with Stephanie Wakefield, Old River Landing, Louisiana, April 14, 2018.

274 Nor is it a matter of proclaiming the unbreakabililty of the poor. For critique of such narratives, often told of the poor or indigenous, in which certain populations are heralded for their supposed talents at coping with and surviving irreparably broken worlds, see David Chandler and Julian Reid, *Becoming Indigenous: Governing Imaginaries in the Anthropocene* (London: Rowman and Littlefield, forthcoming).

275 Isabelle Stengers, *In Catastrophic Times: Resisting the Coming Barbarism* (Open Humanities Press, 2015): 30.

276 Clark, "Volatile Worlds," 2010.

277 Stephen Zavestoski, "Fast Tracking Climate Adaptation: Tapping Our Natural Tendency to Experiment." Our Place on Earth, 2014. https://www.ourplaceonearth.org/blog?category=Guest+Blogger

278 These are questions taken up intellectually as well as practically by Sabu Kohso, in Silvia Federici, George Caffentzis, Anne Waldman, Daniel de Roulet, Sabu Kohso, *Fukushima Mon Amour* (New York: Autonomedia, 2011); Takako Shishido, Ayumi Hirai, Sabu Kohso, Yuko Tonohira, "Voluntary Evacuation: A New Form of Struggle: A Conversation with Takako Shishido by Todos Somos Japon" June 23, 2012. https://jfissures.wordpress.com/2013/01/14/voluntary-evacuation-a-new-form-of-struggle-a-conversation-with-takako-shishido-1/

279 Motonao-gensai Mori, Interview with Stephanie Wakefield, New York, NY, September, 2017.

280 David Kilcullen, *Out of the Mountains: The Coming Age of the Urban Guerrilla* (London: C. Hurst & Co., 2016).

281 Invisible Committee, *To Our Friends* (Los Angeles: Semiotext(e), 2014); Aggregate, *Governing by Design: Architecture, Economy, and Politics in the Twentieth Century* (Pittsburgh: University of Pittsburgh Press, 2012); Michel Foucault, "The Confession of the Flesh."

282 Keller Easterling, *Extrastatecraft: The Power of Infrastructure Space* (New York & London: Verso, 2014), 14.

283 According to the Oxford English Dictionary.

284 NATO Infrastructure Committee, *50 Years of Infrastructure: NATO Security Investment Programme is the Sharing of Roles, Risks, Responsibilities, Costs and Benefits* (Brussels, Belgium: NATO Infrastructure Committee, 2001).

285 Eric Schmidt and Jared Cohen, *The New Digital Age: Transforming Nations, Businesses, and Our Lives* (New York: Vintage Books, 2014), 218.

286 Brian Larkin, "The Politics and Poetics of Infrastructure," *Annual Review of Anthropology*, 42 (2013): 327–343. https://doi.org/ 10.1146/annurev-anthro-092412-155522.

287 Lizabeth Cohen, Lizabeth, *A Consumers' Republic: The Politics of Mass Consumption in Postwar America* (New York: Knopf, 2003).; James T. Patterson, *Grand Expectations the United States, 1945–1974* (Oxford & New York: Oxford University Press, 1996).

288 Colin Waters, Jan Zalasiewicz, Mark Williams, Michael Ellis, and Andrea Snelling, "A Stratigraphical Basis for the Anthropocene?" Geological Society, London, Special Publications, 395 (2014): 1–21. https://doi.org/10.1144/SP395.18.

289 Jan Zalasiewicz, Colin N. Waters, Mark Williams, Anthony D. Barnosky, Alejandro Cearreta, Paul Crutzen, Erle Ellis, Michael A. Ellis, Ian J. Fairchild, Jacques Grinevald, Peter K. Haff, Irka Hajdas, Reinhold Leinfelder, John McNeill, Eric O. Odada, Clement Poirier, Daniel Richter, Will Steffen, Colin Summerhayes, James P.M. Syvitski, Davor Vidas, Michael Wagreich, Scott L. Wing, Alexander P. Wolfe, An Zhishengw, Naomi Oreskes. "When did the Anthropocene begin? A Mid-Twentieth Century Boundary Level is Stratigraphically Optimal." *Quaternary International*, 383 (2015): 196–203. DOI: 10.1016/j.quaint.2014.11.045.

290 Will Steffen, Wendy Broadgate, Lisa Deutsch, Owen Gaffney, and Cornelia Ludwig, "The Trajectory of the Anthropocene: The Great Acceleration." *The Anthropocene Review* 2, no. 1 (2015), 81–98. Doi:10.1177/2053019614564785; Peter K. Haff, "Technology as a Geological Phenomenon: Implications for Human Well Being," Geological Society, London, Special Publications, 395 (2014): 301–309. https://doi.org/10.1144/SP395.4.

291 Christoph Rosol, Sara Nelson, and Jürgen Renn, "In the Machine Room of the Anthropocene," *The Anthropocene Review*, 4, no. 1 (2017): 2–8. https://doi.org/10.1177/2053019617701165.

292 Maria Kaika and Erik Swyngedouw, "Fetishizing the Modern City: The Phantasmagoria of Urban Technological Networks," *International Journal of Urban and Regional Research*, 24, no. 1 (2000): 124. 120–138. https://doi.org/10.1111/1468-2427.00239.

293 Karl Marx and Friedrich Engels, *The German Ideology* (Moscow: Progress Publishers, 1964).

294 Angela Mitropoulos, *Contract & Contagion: From Biopolitics to Oikonomia* (New York: Minor Compositions, 2012), 118.

Notes

295 See e.g. @victoria_falls Instagram account.

296 Greg Glassman, "The World's Most Vexing Problem." *CrossFit*. Video file. September 10, 2017. https://journal.crossfit.com/article/cfj-greg-glassman-the-world-s-most-vexing-problem.

297 Juliet B. Schor, *The Overworked American: The Unexpected Decline in Leisure*4 (New York, NY: Basic Books, 1991), 74.

298 "CrossFit Founder Greg Glassman: 'I'm a rabid libertarian'." ReasonTV. Video File. June 22, 2013. https://www.youtube.com/watch?v=-EB0XyBUl0U

299 Greg Glassman, "CrossFit - The Story of Fran by Greg Glassman." *CrossFit*. Video File. November 28, 2011. https://www.youtube.com/watch?v=-2nsZ9Lbz-8 and Jessie Cameron Herz, *Learning to Breathe Fire: The Rise of CrossFit and the Primal Future of Fitness* (New York: Three Rivers Press, 2014), 20.

300 Glassman, Ibid.

301 Christine Wang, "How a Health Nut Created The World's Biggest Fitness Trend," *CNBC*, April 5, 2016. https://www.cnbc.com/2016/04/05/how-crossfit-rode-a-single-issue-to-world-fitness-domination.html

302 Glassman, "I'm a rabid."

303 Scott Henderson, "CrossFit's Explosive Affiliate Growth by the Numbers," October 23, 2018, https://morningchalkup.com/2018/10/23/crossfits-explosive-affilaite-growth-by-the-numbers/, Julie Jargon, "Too Much Coffee? Starbucks Shops Outnumber McDonald's," *Wall Street Journal*, June 7, 2018, https://www.wsj.com/articles/too-much-coffee-starbucks-shops-outnumber-mcdonalds-1528372800

304 Nellie Bowles, "Exclusive: On the warpath with CrossFit's Greg Glassman," *Maxim*, September 8, 2015, https://www.maxim.com/maxim-man/crossfit-greg-glassman-exclusive-2015-9

305 Agamben, "What is a Destituent."

306 For an excellent critical account of CrossFit contextualized within workplace reorganization and transformation of management techniques post-1970s and contemporary, see Kyle Kubler, "Auto-Body," *Ultra*, August 31, 2016. http://www.ultra-com.org/project/auto-body/.

307 "Steps to Affiliation." *CrossFit*. Accessed March 19, 2019. https://affiliate.crossfit.com/how-to-affiliate

308 Wang, "How a Health Nut."

309 E.g. Eric P. James and Rebecca Gill, "Neoliberalism and the Communicative Labor of CrossFit," *Communication & Sport*, 6, no. 6 (2018): 703–27. Doi:10.1177/2167479517737036, Kubler, 2016. For a similar argument regarding the related Tough Mudder races, see Matthew D. Lamb and Cory Hillman, "Whiners go home: Tough mudder, conspicuous consumption, and the rhetorical proof of 'fitness'," *Communication & Sport* 3, no. 1 (2015): 81–99. Doi:10.1177/2167479514521598.

310 Tyler Graham, "How Kelly Starrett Became CrossFit's Mobility Superhero," *Men's Journal*, 2014, https://www.mensjournal.com/health-fitness/kelly-starrett-interview-crossfit-mobility-mens-fitness/.

311 E.g. Dan Diamond, "Is CrossFit safe? What '60 Minutes' didn't Tell you," *Forbes*, May 11, 2015. https://www.forbes.com/sites/dandiamond/2015/05/11/is-crossfit-good-for-you-what-60-minutes-didnt-say/#4fb9036a508c; Stephanie Cooperman, "Getting Fit Even if it Kills You," *New York Times*, December 22, 2005. https://www.nytimes.com/2005/12/22/fashion/thursdaystyles/getting-fit-even-if-it-kills-you.html.

312 Bowles, "On the Warpath."

313 Kubler, "Auto-Body."

314 Jessie Cameron Herz, "NerdFit: Why Technies Love CrossFit," *BoingBoing*, August 12, 2014. https://boingboing.net/2014/08/12/nerdfit-why-techies-love-cro.html.

315 CrossFit Level 1 Certification Course, author's notes.

316 Glassman, "The World's Most Vexing."

317 Bowles, "On the Warpath."

318 Graham, "How Kelly Starrett."

319 Justin Lofranco, "How Greg Glassman is Reshaping the CrossFit Games," *Morning Chalk Up*, August 23, 2018. https://morningchalk-up.com/2018/08/23/how-greg-glassman-is-reshaping-the-crossfit-games/; Glassman is obsessed with taking down the "Big Sugar" industry and its collaborators in the fitness world. A bit like a serial killer or Fox Mulder, he has a wall devoted to mapping the connections between lobbyists, money into the fitness industry and legislation designed to protect and promote sugar, along with its causing obesity, diabetes and hypertension epidemics, collected under the heading #crushbigsoda. Andréa Maria Cecil, "Big Soda: Buying Chronic Disease." June 26, 2017. *CrossFit Journal*. https://journal.crossfit.com/article/soda-cecil-2017-2; Murray Carpenter, "Mr. CrossFit vs. Big Soda: A profane fitness guru's wonky war

with the soda industry," *Washington Post*, June 4, 2018. https://www.washingtonpost.com/lifestyle/magazine/mr-crossfit-vs-big-soda-a-profane-fitness-gurus-wonky-war-with-the-soda-industry/2018/06/03/88aaa820-5aa6-11e8-8836-a4a123c359ab_story.html?noredirect=on&utm_term=.f0354d61f472.

320 https://crew.endofthreefitness.com/challenge.

321 https://dailystoic.com/.

322 "Ido Portal: Just Move." Brian Rose and London Real. Video File. 2014. https://londonreal.tv/e/ido-portal-just-move/.

323 "Raise Up: The World is Our Gym." Video. Directed by B. Rain Bennett. Flying Flounder Productions, Red Bull Media House. 2017. https://www.redbull.com/int-en/tv/film/AP-1RDEMV9D12111/raise-up.

324 https://dailystoic.com/what-is-stoicism-a-definition-3-stoic-exercises-to-get-you-started/

325 One could also include here Aubrey Marcus, *Own the Day, Own Your Life* (London: Harper Collins, 2018) or the many works of strategy and power writer Robert Greene, such as *The 48 Laws of Power* (New York: Penguin, 1998).

326 Christopher McDougall, *Natural Born Heroes: Mastering the Lost Secrets of Strength and Endurance* (New York: Vintage, 2016), xv; 31.

327 Lanre Bakare, "Chronixx Puts Rastafarianism Back into Jamaican Reggae," *The Guardian*, October 11, 2013, https://www.theguardian.com/music/2013/oct/11/chronixx-roots-reggae, Patricia Meschino, "Is Chronixx reggae's next big thing? Chris Blackwell thinks so!," The Foundation Radio Media Network, September 25, 2013. http://www.clintonlindsay.com/2013/09/25/is-chronixx-reggaes-next-big-thing-chris-blackwell-thinks-so/.

328 Brian Meeks, *Envisioning Caribbean Futures: Jamaican Perspectives* (Kingston: University of the West Indies Press, 2007), 4.

329 Selwyn H.H. Carrington, *The Sugar Industry and the Abolition of the Slave Trade, 1775-1810*, (University Press of Florida, Gainesville, 2002); Robert Roskind, *Rasta heart: A journey into one love* (Blowing Rock, N.C.: One Love Press, 2001); Barry W. Higman, A Concise History of the Caribbean (Cambridge: Cambridge University Press, 2010); Barry W. Higman, *Slave Population and Economy in Jamaica, 1807-1834*. Barbados: The Press University of the West Indies, 1995.

330 Christophe Bonneuil and Jean-Baptiste Fressoz, *The Shock of the Anthropocene: The Earth, History, and Us* (London; Brooklyn, NY: Verso, 2016).

331 David Scott, *Refashioning Futures: Criticism after Postcoloniality*, (Princeton, N.J: Princeton University Press, 1999), 198.

332 David Scott and Stuart Hall, "David Scott by Stuart Hall," *BOMB Magazine*, January 1, 2005, https://bombmagazine.org/articles/david-scott/

333 Scott, *Conscripts of Modernity*, 2004, 2.

334 Hume N. Johnson, "Incivility: The Politics of 'People on the Margins' in Jamaica," *Political Studies* 53, (2005): 579-597. Doi: 10.1111/j.1467-9248.2005.00545.x.

335 Carolyn Cooper, *Noises in the Blood: Orality, Gender, and the "Vulgar" Body of Jamaican Popular Culture* (Durham, N.C.: Duke University Press, 1999); Bakare, "Chronixx Puts Rastafarianism," 2013; Sonjah Stanley Niaah, *Dancehall: From Slave Ship to Ghetto*, (Ottawa: University of Ottawa Press, 2010).

336 Scott and Hall, "David Scott by Stuart Hall," 2005.

337 David Scott, "Political Rationalities of the Jamaican Modern," *Small Axe* 7, no. 2 (2003): 1-22, p. 1.

338 Scott and Hall, "David Scott by Stuart Hall," 2005.

339 Jesse Serwer, "GEN F: Chronixx. Chronixx is leading Jamaican music's roots revival," *THE FADER*, September 20, 2013. https://www.thefader.com/2013/09/20/gen-f-chronixx.

340 Chronixx and Federation Sound, Chronixx Live, The Federation Sound, Red Bull Radio, March 15, 2017, https://soundcloud.com/federationsound/chronixx-live-the-federation-sound-red-bull-radio-031517.

341 Chronixx, Zinc Fence Redemption and Special Guests: A Reggae Jam Session, Concert at Knitting Factory, Brooklyn, NY, May 28, 2018.

342 Ross Kenneth Urken, "What Rasta Taught Chronixx," *FADER*, March 6, 2017, http://www.thefader.com/2017/03/06/what-rasta-taught-chronixx.

343 Ibid.

344 Chronixx, Zinc Fence Redemption.

345 Adidas and &SON, "adidas SPEZIAL Spring Summer '17 featuring Chronixx," Video File. 2017.https://www.youtube.com/watch?v=Ss9sUAr0mc0.

Notes

346 Baz Dreisinger, "It's Not Just Reggae, Says Chronixx: Call It 'Black Experimental Music'," *NPR*, April 21, 2017. https://www.npr.org/templates/transcript/transcript.php?storyId=542628389.

347 David Katz, "Interview: Chronixx," *United Reggae*, July 7, 2017, https://unitedreggae.com/articles/n2206/070717/interview-chronixx-2017.

348 Chronixx, "Jamaican Reggae Artist Chronixx Statement On Vybz Kartel," Accessed March 23, 2019. http://missgaza.com/chronixx-statement-on-vybz-kartel.

349 Proteje featuring Chronixx, "No Guarantee," 2018, from Proteje, *A Matter of Time*, In.Digg.Nation Collective/Overstand Entertainment under license to Easy Star Records, 2018.

350 Chronixx, "On the ends: My Spanish Town by Chronixx," Chronixx.com, May 12, 2016. http://www.chronixxmusic.com/journal-entry/story-of-spanish-town/.

351 Keleshia Powell, "Chronixx Prepared to be Judged for 'Waste Man' Post on Instagram," *Jamaica Observer*, April 10, 2015, http://www.jamaicaobserver.com/news/Chronixx-removes--waste-man--post-from-Instagram-after-harsh-criticisms.

352 Chronixx and Sway, "Chronixx Philosophizes on Death, Rastafarian vs. Religion + Freestyles Live," Sway's Universe, July 13, 2017, https://www.youtube.com/watch?v=PKXEWN-vbEs.

353 Chronixx, Interview. JCTV: Jack Curry TV. Episode 50, Apr 28, 2017, https://www.youtube.com/watch?v=XKMMf9NaMlg.

354 Chronixx, "Here Comes Trouble," on *Dread and Terrible*, Chronixx Music, 2014.

355 Chronixx, "Open Space: Chronixx," *Mass Appeal*, March 24, 2017, http://archive.massappeal.com/open-space-chronixx/.

356 Chronixx and Cipha Sounds, "Chronixx explains why theres no more individuals," Hot 97, September 19, 2013, https://www.youtube.com/watch?v=iqDNXRO8RGE.

357 Chronixx, Zinc Fence Session.

358 See, e.g., "The Productivity-Pay Gap," Economic Policy Institute, 2019, https://www.epi.org/productivity-pay-gap/?fbclid=IwAR364QUg8fcKSTri5pQMzPIA4TUiN0tiQwJUpg1i4tWBuDVj9WiVaPYIn14.

359 The post, worth reading, continues: "I send this message to every youth out there who's wondering why you had to be born into this body and into a world that blames you for everything… I am not

saying this as a perfect man, not as no great sage that know everything but as your young brother who has discovered that there is an abundance of peace in a heart that forgives. learn to forgive! Forgive yourself because none of this is your fault. Forgive yourself for all the hate you have been made to project into your own reality. Trust your most high self and forgive the generations before and behind you because it is us the living youths of the future who have the power to purify the energy they left in the Earth. We never really die...it is us crying out from our past to our present manifestation for forgiveness and healing. There is no way a 90s baby like myself could be responsible for these 500 year old problems that are snatching our youths life away before they are even aware of what this opportunity to live is really about. Still we have the power to heal ourselves if we wish. The one thing I imagined that could be worst than being murdered is to have your mind programmed to hate the body you were born in so much that you resent it....love your body. No matter how your temple looks...it is in it that your soul chose to work...have reverence!and have love!...I have a ever growing love and a great hope for all my talented sisters and brothers out there 25yo and younger. Stay strong! A child shall lead them—Jamar." Chronixx (@chronixxmusic). 2018. "Us the youths of today have found ourselves." Instagram, June 18, 2018. https://www.instagram.com/p/BkMAtoxF6Jl/?utm_source=ig_share_sheet&igshid=1j3hdu0wc88kz

360 Chronixx and Federation Sound, "Chronixx Live."

361 Andrew Revkin, "Varied Views (Dark, Light, in Between) of Earth's Anthropocene Age," Dot Earth New York Times blog, July 15, 2015, https://dotearth.blogs.nytimes.com/2015/07/15/varied-views-dark-light-in-between-of-earths-anthropocene-age. Those for "good" Anthropocenes argue for planetary management, a la Rockström, imagine a new stage where humans become Earth systems "stewards" or deploy accelerated technology (nuclear, genetic engineering, geoengineering, carbon capture, industrial agriculture). On the latter, see John Asafu-Adjaye, Blomqvist, Linus, Brand, Stewart, Brook, Barry, DeFries, Ruth, Ellis, Erle, Foreman, Christopher, Keith, David, Lewis, Martin, Lynas, Mark, Nordhaus, Ted, Pielke, Roger, Pritzker, Rachel, Roy, Joyashree, Sagoff, Mark, Shellenberger, Michael, Stone, Robert, Teague, Peter, "An Ecomodernist Manifesto," Breakthrough Institute, 2015. http://www.ecomodernism.org/. On "bad" Anthropocene see Clive Hamilton, "The Delusion of the "Good Anthropocene": Reply to Andrew Revkin," June 17, 2014, https://clivehamilton.com/the-delusion-of-the-good-anthropocene-reply-to-andrew-revkin/. To be clear I am not arguing for either of these sides. As always, what matters is the third term—less quantifiable, more complex—which is repeatedly denied by such discourse.

362 Nassim Nicholas Taleb, *Skin in the game: Hidden Asymmetries in Daily Life* (New York: Random House, 2018).

363 Such back loop practices occur within and reclaim the "implicated," multiple rhythms of everyday time which Sébastian Norbert et al (2017) argue are entrapped by resilience's homogenous, future-oriented temporality. Sébastian Norbert, "Resilience for the Anthropocene? Shedding Light on The Forgotten Temporalities Shaping Post-Crisis Management in the French Sud Ouest," *Resilience* 5, no. 3 (2017): 145–160, Doi: 10.1080/21693293.2016.1241479.

364 Supreme Understanding, *Knowledge of Self*, 8.

365 This is the view forwarded at times by Michel de Certeau's otherwise excellent development of the concept of use. See Michel de Certeau, *The Practice of Everyday Life* (Berkeley: University of California Press, 1988). See also AbdouMaliq Simone, *For the City Yet to Come: Changing African Life in Four Cities* (Durham: Duke University Press, 2004).

366 Foucault, *The Use of Pleasure*; Giorgio Agamben. "What is a Destituent Power?" *Environment and Planning D: Society and Space* 32, no. 1 (2014): 65–74; Giorigo Agamben. *The Use of Bodies* (Stanford: Stanford University Press, 2016).

367 Martin Heidegger, *Being and Time*, Trans. John Macquarrie and Edward Robinson (New York: Harper & Row, 1963), 169-224.

368 Foucault, *The Use of Pleasure*.

369 Of course these practices were devised by and for free adult men. Foucault, *The Use of Pleasure*, 47. But the point is not the Greeks, but what this concept of use can help us see and do today.

370 For a different version of this argument, see Bruce Braun and Stephanie Wakefield. "Destituent Power and Common Use: Reading Agamben in the Anthropocene," in *Handbook on the Geographies of Power*, ed. Mat Coleman and John Agnew (Georgia: University of Georgia Press, 2018), 259-276.

371 Timothy Morton, *Hyperobjects: Philosophy and Ecology After the End of the World* (Minneapolis: University of Minnesota Press, 2013). Thank you to Andrew Baldwin for pointing out this connection.

372 Foucault, *The Use of Pleasure*, 40. Likewise philosopher Emmanuel Levinas argued that one does not first and foremost use a tool for practical purposes but for enjoyment. "They are always in a certain measure—and even hammers, needles, and machines are—objects of enjoyment, presenting themselves to 'taste,' already adorned, embellished." Emmanuel Levinas, *Totality and Infinity: An Essay*

in *Exteriority* (Pittsburgh, Pennsylvania: Duquesne University Press, 1969), 110.

373 Elizabeth Grosz, *Chaos, Territory, Art: Deleuze and the Framing of the Earth* (Durham, N.C.: Duke University Press, 2008), 2.

374 Adidas and &SON, "adidas SPEZIAL."

375 Giorgio Agamben, *The Use of Bodies* (Stanford: Stanford University Press, 2016), 220.

376 On practices of liberty see Michel Foucault, "The Ethic of the Care for the Self as a Practice of Freedom: An Interview with Michael Foucault on 20th January 1984," in *The Final Foucault*, edited by James William Bernauer and David M. Rasmussen (Cambridge, MA: MIT Press, 1987).

377 From a version of the story for children: James Baldwin, *Old Greek Stories* (Fairfield, IA: 1st World Library Literary Society, 2004).

378 Aeschylus, trans. Weir Smyth, *Prometheus Bound*, (Cambridge: Harvard University Press, 1926). Apollonius Rhodius, trans. E. V. Rieu, *Argonautica* (London: Penguin Classics, 1959/1971).

379 For an interesting recent development of the concept of technics see Yuk Hui, *The Question Concerning Technology in China: An Essay in Cosmotechnics* (Falmouth, UK: Urbanomic Media Ltd, 2016).

380 Nigel Clark, "Metamorphoses: on Philip Conway's Geopolitical Latour," *Global Discourse: An Interdisciplinary Journal of Current Affairs and Applied Contemporary Thought* 6, no. 1-2 (2016): 73. Doi: 10.1080/23269995.2015.1062294.

381 Gilles Deleuze and Claire Parnet, *Gilles Deleuze From A to Z*. "G for Gauche." DVD. (Semiotext(e), 2011).

382 Donald Trump (@realDonaldTrump), 2017. "Writing my inaugural address at the Winter White House, Mar-a-Lago, three weeks ago. Looking forward to Friday. #Inauguration." Twitter, January 18, 2017, 12:33 PM. https://twitter.com/realDonaldTrump/status/821772494864580614.

383 Apollodorus, *Gods and Heroes of the Greeks: The Library of Apollodorus* (Amherst, MA: University of Massachusetts Press, 1976), 127; Rhys Carpenter, *Beyond the Pillars of Hercules: The Classical World Seen Through the Eyes of its Discoverers* (New York, NY: Delacorte Press, 1966).

384 Peter Sloterdijk, *You Must Change Your Life* (Cambridge, UK: Polity Press, 2013).

385 Sloterdijk, *You Must Change*, 14.

386 Nigel Clark, *Inhuman Nature: Sociable Life on a Dynamic Planet* (London: Sage, 2011).

387 Nigel Clark, Alexandra Gormally, and Hugh Tuffen, "Speculative Volcanology: Time, Becoming and Violence in Encounters with Magma," *Environmental Humanities* 10, no. 1 (2018): 289. 273-294 Doi: 10.1215/22011919-4385571.

388 Clark, "Volatile Worlds," 11.

389 Clark, Ibid, 12.

390 Jamie Lorimer, *Wildlife in the Anthropocene: Conservation after Nature* (Minneapolis: University of Minnesota Press, 2015); Jamie Lorimer and Clemens Driessen, "Wild Experiments at the Oostvaardersplassen: Rethinking Environmentalism in the Anthropocene," *Transactions of the Institute of British Geographers* 39, no. 2 (2013): 169–181. Doi:10.1111/tran.12030.

391 Jamie Lorimer, Chris Sandom, Paul Jepson, Chris Doughty, Maan Barua, and Keith J. Kirby, "Rewilding: Science, Practice, and Politics," *Annual Review of Environment and Resources* 40, (2015): 39–6. Doi: 10.1146/annurev-environ-102014-021406.

392 Consider the recent catastrophic failure of the Dutch experiment in rewilding a section of land near Amsterdam, at Oostvaardersplassen. Patrick Barkham, "Dutch Rewilding Experiment Sparks Backlash as Thousands of Animals Starve," *The Guardian*. April 27, 2018. https://www.theguardian.com/environment/2018/apr/27/dutch-rewilding-experiment-backfires-as-thousands-of-animals-starve.

393 Martin Savransky, *The Adventure of Relevance: An Ethics of Social Inquiry* (New York, NY: Palgrave Macmillan, 2016).

394 "Billions of back loops" is a reference to Kathryn Yusoff, *A Billion Black Anthropocenes or None* (Minneapolis: University of Minnesota Press, 2019).

395 Sebastian Junger, *Tribe: On Homecoming and Belonging* (New York: Hachette Book Group, 2016); David Ronfelt, "IN SEARCH OF HOW SOCIETIES WORK: Tribes: The First and Forever Form." RAND Corporation Working Paper WR-433-RPC, 2007; Limberg and Barnes, "Mimetic Tribes"; Roger Burrows, "Urban Futures and The Dark Enlightenment: A Brief Guide for the Perplexed," in *Towards a Philosophy of the City: Interdisciplinary and Transcultural Perspectives*, edited by Keith Jacobs and Jeff Malpas (London: Rowman and Littlefield, 2018).

396 Noortje Marres, "Why We Can't Have Our Facts Back," *Engaging Science, Technology, and Society*, 4 (2018): 423-443. Doi:10.17351/ests2018.188; Benjamin Neimark, John Childs, Andrea J.

Nightingale, Connor Joseph Cavanagh, Sian Sullivan, Tor A. Benjaminsen, Simon Batterbury, Stasja Koot, and Wendy Harcourt, "Speaking Power to 'Post-Truth': Critical Political Ecology and the New Authoritarianism," *Annals of the American Association of Geographers* 109, no. 2 (2019): 613-623.

397 Phil Neel, *Hinterland: America's New Landscape of Class and Conflict* (London: Reaktion Books, 2018).

398 David Kilcullen, *Out of the Mountains: The Coming Age of the Urban Guerilla* (Oxford: Oxford University Press, 2013).

399 Bruce Elliott Johansen and Barbara Alice Mann, eds., *Encyclopedia of the Haudenosaunee (Iroquois Confederacy)* (Westport, CT: Greenwood Publishing Group, 2000). On Ganienkeh, see "They Call Us Mohawks: Ganienkeh: A Way of Life." Turtle Island Trust Documentary, 2013. https://www.youtube.com/watch?v=9NqeyFIWnlc.

400 Author's notes.

401 Jean Baudrillard, *Seduction* (Montréal: New World Perspectives, 1990).

Works Cited

Adams, Ross Exo. "Notes from the Resilient City." *Log*, 32 (2014).

Adams, Ross Exo. "Becoming-Infrastructural." *e-flux architecture*, 2017. http://www.e-flux.com/architecture/positions/149606/becoming-infrastructural/.

Adragna, Anthony. "Florida Senator Nelson calls sea-level rise, climate change 'compelling story,' announces hearing." *Bloomberg BNA*, March 12, 2014. https://www.bna.com/florida-sen-nelson-b17179882772/.

Aeschylus. *Prometheus Bound*. Translated by Weir Smyth. Cambridge: Harvard University Press, 1926.

Agamben, Giorgio. *The Time That Remains*. Stanford: Stanford University Press, 2005.

Agamben, Giorgio. "What is a Destituent Power?" *Environment and Planning D: Society and Space* 32, no. 1 (2014): 65–74.

Agamben, Giorgio. *The Use of Bodies*. Stanford: Stanford University Press, 2016.

Aggregate. *Governing by Design: Architecture, Economy, and Politics in the Twentieth Century*. Pittsburgh: University of Pittsburgh Press, 2012.

Anthropocene Working Group, Media note. August 29, 2016. http://www2.le.ac.uk/offices/press/press-releases/2016/august/media-note-anthropocene-working-group-awg.

Apollodorus. *Gods and Heroes of the Greeks: The Library of Apollodorus*. Translated by Michael Simpson. Amherst, MA: University of Massachusetts Press, 1976.

Apollonius Rhodius. *Argonautica*. Translated by E. V. Rieu. London: Penguin Classics, 1959/1971.

Asafu-Adjaye, John, Blomqvist, Linus, Brand, Stewart, Brook, Barry, DeFries, Ruth, Ellis, Erle, Foreman, Christopher, Keith, David, Lewis, Martin, Lynas, Mark, Nordhaus, Ted, Pielke, Roger, Pritzker, Rachel, Roy, Joyashree, Sagoff, Mark, Shellenberger, Michael, Stone, Robert, Teague, Peter, "An Ecomodernist Manifesto." Breakthrough Institute, 2015. http://www.ecomodernism.org/.

Australian Government and Queensland Government. "Highlights of the Reef 2050 Long-Term Sustainability Plan." 2015. http://www.environment.gov.au/marine/gbr/publications/highlights-long-term-sustainability-plan#.

Bakare, Lanre. "Chronixx Puts Rastafarianism Back into Jamaican Reggae." *The Guardian*, October 11, 2013. https://www.theguardian.com/music/2013/oct/11/chronixx-roots-reggae.

Baldwin, James. *Old Greek Stories*. Fairfield, IA: 1st World Library Literary Society, 2004.

Barkham, Patrick. "Dutch Rewilding Experiment Sparks Backlash as Thousands of Animals Starve." *The Guardian*. April 27, 2018. https://www.theguardian.com/environment/2018/apr/27/dutch-rewilding-experiment-backfires-as-thousands-of-animals-starve.

Barnett, Clive and Gary Bridge. "Thinking Problematically About the City." *International Journal of Urban and Regional Research* 40, 6 (2016): 1186–1204.

Baudrillard, Jean. *Seduction*. Montréal: New World Perspectives, 1990.

Beckett, Katherine and Steve Herbert. "Dealing with disorder: Social control in the post-industrial city." *Theoretical Criminology* 12, no. 1 (2008): 5-30.

Benanav, Aaron and John Clegg. "Misery and Debt: On the Logic and History of Surplus Populations and Surplus Capital." *Endnotes* 2, 2010. https://endnotes.org.uk/issues/2/en/endnotes-misery-and-debt.

Benanav, Aaron. *A Global History of Unemployment since 1949*. London: Verso, forthcoming.

Benjamin, Walter. *Illuminations*. New York: Shocken Books, 1968.

Bennett, Jane. *Vibrant Matter: A Political Ecology of Things*. Durham: Duke University Press, 2007.

Bennett, Jessica. "Rise of the Preppers: America's New Survivalists." *Newsweek*, December 29, 2009. https://www.newsweek.com/rise-preppers-americas-new-survivalists-75537.

Berkes, Fikret and Carl Folke, eds. *Linking Social and Ecological Systems: Management Practices and Social Mechanisms for Building Resilience*. Cambridge University Press, New York, 1998.

Berkes, Fikret, Johan Colding, Carl Folke, eds. *Navigating Social-Ecological Systems: Building Resilience for Complexity and Change*. Cambridge: Cambridge University Press, 2003.

Berkowitz, Michael. "The Movement We're Building." *100 Resilient Cities*, July 24, 2017. https://www.100resilientcities.org/the-movement-were-building/. On 100 Resilient Cities, see https://www.100resilientcities.org/about-us/#section-2.

Berlant, Lauren. *Cruel Optimism*. Durham, NC: Duke University Press, 2011.

Berlant, Lauren. "The Commons: Infrastructures for Troubling Times." *Environment and Planning D: Society and Space* 34, no. 3 (2016): 393-419. Doi: 10.1177/0263775816645989.

Beuret, Nicholas and Gareth Brown. "The Walking Dead: The Anthropocene as a Ruined Earth." *Science as Culture* 26, no. 3 (2017): 330–354. Doi: 10.1080/09505431.2016.1257600.

BIG Bjarke Ingels Group. "The Big U." Final proposal boards submitted to Rebuild by Design. 2014. https://www.dropbox.com/s/enf240jnzyzoftw/ BIG_RBD_131028_All%20Boards.pdf.

Bloomberg, Michael. Mayor Bloomberg Delivers Address on Shaping New York City's Future after Hurricane Sandy. Press conference, New York, Marriot Downtown, December 6, 2012. http://www.nyc.gov/portal/site/nycgov/menuitem.c0935b9a57b-b4ef3daf2f1c701c789a0/index.jsp?pageID=mayor_press_release&catID=1194&doc_name=http://www.nyc.gov/html/om/html/2012b/pr459-12.html&cc=unused1978&rc=1194&ndi=1.

Bloomberg, Michael. "Mayor Bloomberg Presents the City's Longterm Plan to Further Prepare for the Impacts of a Changing Climate." Press conference, New York, June 11, 2013. http://www1.nyc.gov/office-of-the-mayor/news/200-13/mayor-bloomberg-presents-city-s-long-term-plan-further-prepare-the-impacts-a-changing.

Bonneuil, Christophe and Jean-Baptiste Fressoz. *The Shock of the Anthropocene: The Earth, History, and Us*. London; Brooklyn, NY: Verso, 2016.

Bowles, Nellie. "Exclusive: On the Warpath with CrossFit's Greg Glassman." *Maxim*, September 8, 2015. https://www.maxim.com/maxim-man/crossfit-greg-glassman-exclusive-2015-9.

Boyer, Christine. *Dreaming the Rational City: The Myth of American City Planning*. Cambridge: MIT Press, 1986.

Braun, Bruce and Noel Castree, *Social Nature: Theory, Practice and Politics*. London: Wiley- Blackwell, 2001.

Braun, Braun. "A New Urban Dispositif? Governing Life in an Age of Climate Change." *Environment and Planning D: Society and Space* 32, no. 1 (2014): 49-64.

Braun, Bruce and Stephanie Wakefield. "Destituent Power and Common Use: Reading Agamben in the Anthropocene." In *Handbook on the Geographies of Power*, edited by Mat Coleman and John Agnew 259-276.Georgia: University of Georgia Press, 2018. pp. 259-276.

Brecher, Jeremy. *Strike!* Cambridge, MA: South End Press, 1997.

Bulkeley, Harriet, Simon Marvin, Yuliya Voytenko Palgan, Kes McCormick, Marija Breitfuss-Loidl, Lindsay Mai, Timo von Wirth, and Niki Frantzeskaki. "Urban Living Laboratories: Conducting the Experimental City?" *European Urban and Regional Studies*, 2018. Doi:10.1177/0969776418787222

Bulkeley, Harriet, Vanesa Castán Broto, and Gareth A.S. Edwards. *An Urban Politics of Climate Change: Experimentation and the Governing of Socio-Technical Transitions*. London: Routledge, 2015.

Burrows, Roger. "Urban Futures and The Dark Enlightenment: A Brief Guide for the Perplexed." In *Towards a Philosophy of the City: Interdisciplinary and Transcultural Perspectives*, edited by Keith Jacobs and Jeff Malpas, 245-258. London: Rowman and Littlefield, 2018.

Carpenter, Rhys. *Beyond the Pillars of Hercules: The Classical World Seen Through the Eyes Of its Discoverers*. New York, NY: Delacorte Press, 1966.

Carrington, Selwyn H. H. *The Sugar Industry and the Abolition of the Slave Trade, 1775-1810*. Gainesville, FL: University Press of Florida, 2002.

Carse, Ashley. "Keyword: Infrastructure: How A Humble French Engineering Term Shaped The Modern World." In *Infrastructures and Social Complexity: A Companion*, edited by Penelope Harvey, Casper Bruun Jensen, Atsuro Morita, 27-39. Abingdon: Routledge, 2017.

Castells, Manuel. *Networks of Outrage and Hope: Social Movements in the Internet Age*. Cambridge: Polity, 2015.

Cavanaugh, Kyle C. "Poleward Expansion of Mangroves is a Threshold Response to Decreased Frequency of Extreme Cold Events." *Proceedings of the National Academy of Sciences* 111, no. 2 (2014): 723—727.

Cavelty, Myriam Dunn and Kristian Soby Kristensen. *Securing "the Homeland": Critical Infrastructure, Risk and (In)Security*. London and New York: Routledge, 2008.

Cecil, Andréa Maria. "Big Soda: Buying Chronic Disease." June 26, 2017. *CrossFit Journal.* https://journal.crossfit.com/article/soda-cecil-2017-2

Chandler, David. *Resilience: The Governance of Complexity*. Abingdon: Routledge, 2014.

Chandler, David and Julian Reid, *The Neoliberal Subject: Resilience, Adaptation and Vulnerability*. Rowman & Littlefield International, London, 2016.

Chandler, David and Julian Reid, *Becoming Indigenous: Governing Imaginaries in the Anthropocene*. London: Rowman and Littlefield, forthcoming.

Chen, I-Ching, Jane K. Hill, Ralf Ohlemüller, David B. Roy, Chris D. Thomas. "Rapid Range Shifts of Species Associated with High Levels of Climate Warming." *Science* 333, no. 6045 (2011): 1024—1026.

Chronixx and Cipha Sounds. "Chronixx Explains Why There's No More Individuals." *Hot 97*, September 19, 2013. https://www.youtube.com/watch?v=iqDNXRO8RGE.

Chronixx. "Here Comes Trouble." *Dread and Terrible*. Chronixx Music, 2014.

Chronixx and Federation Sound. Chronixx Live. The Federation Sound, Red Bull Radio, March 15, 2017. https://soundcloud.com/federationsound/chronixx-live-the-federation-sound-red-bull-radio-031517.

Chronixx and Sway. "Chronixx Philosophizes on Death, Rastafarian vs. Religion + Freestyles Live." *Sway's Universe*, July 13, 2017. https://www.youtube.com/watch?v=PKXEWN-vbEs.

Chronixx. "I Can." From *Chronology*. Soul Circle Music/Virgin EMI Records, 2017.

Chronixx. "Open Space: Chronixx." *Mass Appeal,* March 24, 2017. http://archive.massappeal.com/open-space-chronixx/.

Chronixx. Interview. JCTV: Jack Curry TV. Episode 50, Apr 28, 2017. https://www.youtube.com/watch?v=XKMMf9NaMlg.

Chronixx (@chronixxmusic). 2018. "Us the youths of today have found ourselves." Instagram, June 18, 2018. https://www.instagram.com/p/BkMAtoxF6Jl/?utm_source=ig_share_sheet&igshid=1j3hdu0wc88kz.

Citarella, Joshua. "Politigram and the post-Left, short version." 2018. http://joshuacitarella.com/_pdf/Politigram_Post-left_2018_short.pdf.

City of New York. "PlaNYC: A stronger, more resilient New York." Report prepared by the Special Initiative for Rebuilding and Resiliency. 2013, 5-6. http://www.nyc.gov/html/sirr/html/report/report.shtml.

City of New York. "Mayor Bloomberg, police commissioner Kelly and Microsoft unveil new, state-of-the-art law enforcement technology that aggregates and analyzes existing public safety data in real time to provide a comprehensive view of potential threats and criminal activity." Press release, New York, August 8, 2012. http://www.nyc.gov/portal/site/nycgov/menuitem.c0935b9a57b-b4ef3daf2f1c701c789a0/index.jsp?pageID=mayor_press_release&catID=1194&doc_name=http://www.nyc.gov/html/om/html/2012b/pr291-12.html&cc=unused1978&rc=1194&ndi=1

Clark, Nigel. "Volatile Worlds, Vulnerable Bodies Confronting Abrupt Climate Change." *Theory, Culture & Society* 27, no. 2-3 (2010): 31-53. Doi: 10.1177/0263276409356000.

Clark, Nigel. *Inhuman Nature: Sociable Life on a Dynamic Planet* London: Sage, 2011.

Clark, Nigel. "Metamorphoses: On Philip Conway's Geopolitical Latour." *Global Discourse:*

An Interdisciplinary Journal of Current Affairs and Applied Contemporary Thought 6, no. 1-2 (2016): 72-75. Doi: 10.1080/23269995.2015.1062294.

Clark, Nigel and Kathryn Yusoff. "Geosocial Formations and the Anthropocene." *Theory, Culture & Society* 34, no. 2-3 (2017): 3–23. Doi:10.1177/0263276416688946.

Clark, Nigel, Alexandra Gormally, and Hugh Tuffen. "Speculative Volcanology: Time, Becoming and Violence in Encounters with Magma." *Environmental Humanities* 10, no. 1 (2018): 273-294. Doi: 10.1215/22011919-4385571.

Coaffee, Jon. "Risk, Resilience and Environmentally Sustainable Cities." *Energy Policy* 36, no. 12, (2008), 4633-4638: 4633.

Coaffee, Jon and Peter Lee. *Urban Resilience: Planning for Risk, Crisis, and Uncertainty*. New York and London: Palgrave, 2017.

Cohen, Lizabeth. *A Consumers' Republic: The Politics of Mass Consumption in Postwar America*. New York: Knopf, 2003.

Colebrook, Claire and Jami Weinstein. "Preface: Postscript on the posthuman." In *Posthumous Life: Theorizing Beyond the Posthuman*, edited by Jami Weinstein and Claire Colebrook, ix-xxix. New York, NY: Columbia University Press, 2017.

Cooper, Carolyn. *Noises in the Blood: Orality, Gender, and the "Vulgar" Body of Jamaican Popular Culture*. Durham, N.C.: Duke University Press, 1999.

Cooperman, Stephanie. "Getting Fit Even if it Kills You." *New York Times*, December 22, 2005. https://www.nytimes.com/2005/12/22/fashion/thursdaystyles/getting-fit-even-if-it-kills-you.html.

Copeland, Baden, Josh Keller, and Bill Marsh. "What Could Disappear." *New York Times*, November 24, 2012. http://www.nytimes.com/interactive/2012/11/24/opinion/sunday/ what-could-disappear.html.

Cowie, Jefferson. *Stayin' Alive: The 1970s and the Last Days of the Working Class*. New York: The New Press, 2010.

Crist, Eileen. "On the Poverty of our Nomenclature." *Environmental Humanities* 3, no. 1 (2013): 129-147. Doi: 10.1215/22011919-3611266

CrossFit. "Steps to Affiliation." *CrossFit*. Accessed March 19, 2019. https://affiliate.crossfit.com/how-to-affiliate.

Crowl, Todd A. and Rita A. Teutonico. "As the Sea Rises, South Floridians Will Get Thirsty Before They Get Wet." *Sun Sentinel*, June 4, 2018. https://www.sun-sentinel.com/opinion/commentary/fl-op-viewpoint-sea-level-rise-freshwater-supply-20180601-story.html.

Crutzen, Paul. "Geology of Mankind." *Nature* 415, no. 6867 (2002): 23. Doi:10.1038/415023a.

Crutzen, Paul and Eugene Stoermer. "Have We Entered the 'Anthropocene'?" *International Geosphere-Biosphere Programme (IGBP) Newsletter* 41 (2000): 17–18.

Cuomo, Andrew. "We Will Lead on Climate Change." *New York Daily News*, November 25, 2012. http://www.nydailynews.com/opinion/lead-climate-change-article-1.1202221.

Danowski, Déborah and Eduardo Viveiros de Castro. *The Ends of the World*. Cambridge, UK: Polity Press, 2017.

Davenport, Coral and Campbell Robertson. "Resettling the First American 'Climate Refugees'." *New York Times*, May 3, 2016. https://www.nytimes.com/2016/05/03/us/resettling-the-first-american-climate-refugees.html.

Davis, Mike. *Ecology of Fear: Los Angeles and the Imagination of Disaster*. New York: Metropolitan Books, 1998.

De Certeau, Michel. *The Practice of Everyday Life*. Berkeley: University of California Press, 1988.

De la Cadena, Marisol and Mario Blaser. *A World of Many Worlds*. Durham, NC: Duke University Press, 2018.

Deering Estate. "Rehydration Project." 2019. https://deeringestate.org/conservation/rehydration-project/.

Deleuze, Gilles and Claire Parnet. *Gilles Deleuze From A to Z*. DVD. New York: Semiotext(e), 2011.

Diamond, Dan. "Is CrossFit safe? What '60 Minutes' Didn't Tell You." *Forbes*, May 11, 2015. https://www.forbes.com/sites/dandiamond/2015/05/11/is-crossfit-good-for-you-what-60-minutes-didnt-say/#4fb9036a508c.

Dillon, Michael. "Governing Terror: The State of Emergency of Biopolitical Emergence." *International Political Sociology* 1 no. 1 (2007): 7-28.

Dillon, Michael and Julian Reid. *The Liberal Way of War: Killing to Make Life Live*. London: Routledge, 2009.

Eakin, Emily. "The Civilization Kit." *The New Yorker*, December 23 and 30, 2013. https://www.newyorker.com/magazine/2013/12/23/the-civilization-kit.

Easterling, Keller. *Extrastatecraft: The Power of Infrastructure Space*. New York & London: Verso, 2014.

Economist. "Preparing for the Apocalypse: I Will Survive." *Economist*, December 17, 2014. https://www.economist.com/christmas-specials/2014/12/17/i-will-survive

Eisinger, Peter. "The Politics of Bread and Circuses: Building the City for the Visitor Class." *Urban Affairs Review* 35, no. 3 (2000): 316-333. Doi: 10.1177/107808740003500302.

Escobar, Arturo. *Designs for the pluriverse: Radical Interdependence, Autonomy, and the Making of Worlds*. Durham, N.C.: Duke University Press, 2018.

Evans, Brad and Julian Reid. *Resilient Life: The Art of Living Dangerously*. Cambridge, UK: Polity, 2014.

Evans, James and Andrew Karvonen. "Give Me a Laboratory and I Will Lower Your Carbon Footprint!'—Urban Laboratories and the Governance of Low-Carbon Futures." *International Journal of Urban and Regional Research* 38 no. 2 (2014): 413-430.

Evans, Joshua. "Trials and Tribulations: Conceptualizing the City through/as Urban Experimentation." *Geography Compass* 10, no. 10 (2016): 429–443.

Fagan, Brian. *The Long Summer: How Climate Changed Civilization*. London: Granta Books, 2004.

Fath, Brian D., Carly A. Dean, and Harald Katzmair. "Navigating the Adaptive Cycle: An Approach to Managing the Resilience of Social Systems." *Ecology and Society* 20, no. 2 (2015): 24. Doi:10.5751/ES-07467-200224.

Federal Emergency Management Association (FEMA). "A Whole Community Approach to Emergency Management: Principles, Themes, and Pathways for Action." Washington, DC: US Department of Homeland Security, 2011. https://www.fema.gov/media-library-data/20130726-1813-25045-0649/whole_community_dec2011__2_.pdf.

Finney, Stanley and Lucy Edwards. "The "Anthropocene" Epoch: Scientific Decision or Political Statement?" *GSA Today* 26, no. 3 (2016): 614-621.

Flechas, Joey and Jenny Staletovich. "Miami's battle to stem rising tides." *Miami Herald*, October 23, 2015. https://www.miamiherald.com/news/local/community/miami-dade/miami-beach/article41141856.html.

Flood, Alison. "'Post-truth' Named Word of the Year by Oxford Dictionaries." *The Guardian.* November 15, 2016. https://www.theguardian.com/books/2016/nov/15/post-truth-named-word-of-the-year-by-oxford-dictionaries.

Florida, Richard and Sara Johnson. "Making Our Coastal Cities More Resilient Can't Wait." *City Lab,* November 1, 2012. http://www.citylab.com/work/2012/11/ making-our-cities-more-resilient-cant-wait/3758/.

Florida, Richard and Andrew Zolli. "The Rush to Resilience: We Don't Have Decades Before the Next Sandy." *City Lab,* November 9, 2012. http://www.citylab.com/work/2012/11/building-resilient-cities-conversation-andrew-zolli-and-jonathan-rose/3839/.

Folger, Tim. "Rising Seas." *National Geographic,* September, 2013. http://ngm.nationalgeographic.com/2013/09/rising-seas/folger-text.

Folke, Carl. "Resilience: The Emergence of a Perspective for Social–Ecological Systems Analyses." *Global Environmental Change* 16 (2006): 253–267.

Folke, Carl, Stephen R. Carpenter, Brian Walker, Marten Scheffer, Terry Chapin, and Johan Rockström. "Resilience Thinking: Integrating Resilience, Adaptability and Transformability." *Ecology and Society* 15, no. 4 (2010): 20. http://www.ecologyandsociety.org/vol15/iss4/art20/.

Foucault, Michel. *Discipline and Punish: The Birth of the Prison.* New York: Pantheon Books, 1977.

Foucault, Foucault. "The Confession of the Flesh." In *Power/Knowledge: Selected Interviews and Other Writings,* edited by Colin Gordon, 194-228. New York: Pantheon Books, 1980.

Foucault, Michel. "Truth and Power." In *The Foucault Reader,* edited by Paul Rabinow, 51-75. New York: Pantheon Books, 1984.

Foucault, Michel. "The Ethic of the Care for the Self as a Practice of Freedom: An Interview with Michael Foucault on 20th January 1984." In *The Final Foucault,* edited by James William Bernauer and David M. Rasmussen. Cambridge, MA: MIT Press, 1987.

Foucault, Michel. *The History of Sexuality, Volume 2: The Use of Pleasure.* New York: Vintage, 1986.

Foucault, Michel. *The History of Sexuality, Volume 3: The Care of the Self.* New York: Vintage, 1988.

Foucault, Michel. *The Government of Self and Others: Lectures at the Collège de France 1982–1983*. New York: Palgrave Macmillan, 2010.

Fukuyama, Francis. *The End of History and the Last Man*. New York: The Free Press, 1992.

Furloni, Mario. "The (Open-Source) Ecologist." Video file. Accessed March 19, 2019. https://vimeo.com/26613005.

Gaffney, Owen. "Walking the Anthropocene." *National Geographic*, 2015, March 16. https://www.nationalgeographic.org/projects/out-of-eden-walk/blogs/lab-talk/2015-03-walking-anthropocene/.

Gaffney, Owen, and Will Steffen. "The Anthropocene Equation." *The Anthropocene Review* 4, no. 1 (2017): 53–61.

Gandy, Matthew. *Concrete and Clay: Reworking Nature on New York City*. Cambridge, MA: MIT Press, 2003.

Glaberman, Martin. *Punching Out & Other Writings*. Chicago: Charles H. Kerr Publishing Company, 2002.

Glassman, Greg. "CrossFit - The Story of Fran by Greg Glassman." CrossFit. Video File. November 28, 2011. https://www.youtube.com/watch?v=-2nsZ9Lbz-8

Glassman, Greg. "CrossFit Founder Greg Glassman: 'I'm a rabid libertarian'." *ReasonTV*. Video File. June 22, 2013. https://www.youtube.com/watch?v=-EB0XyBUl0U.

Glassman, Greg. "The World's Most Vexing Problem." *CrossFit*. Video file. September 10, 2017. https://journal.crossfit.com/article/cfj-greg-glassman-the-world-s-most-vexing-problem.

Goodell, Jeff. "Miami: How Rising Sea Levels Endanger South Florida." *Rolling Stone*, June 20, 2013. https://www.rollingstone.com/politics/politics-news/miami-how-rising-sea-levels-endanger-south-florida-200956/.

Graham, Stephen. "Urban Metabolism as Target: Contemporary War as Forced Demodernization." In *The Nature of Cities: Urban Political Ecology and the Politics of Urban Metabolism*, edited by Nik Heynen, Maria Kaika, and Erik Swyngedouw, 234-252, London: Routledge, 2006.

Graham, Stephen. *Cities Under Siege: The New Military Urbanism*. New York and London: Verso, 2011.

Graham, Tyler. "How Kelly Starrett Became CrossFit's Mobility Superhero." *Men's Journal,* 2014. https://www.mensjournal.com/health-fitness/kelly-starrett-interview-crossfit-mobility-mens-fitness/

Greater Miami and the Beaches. "Preliminary Resilience Assessment." 100 Resilient Cities, report, 2017. http://www.mbrisingabove.com/wp-content/uploads/2017/10/170905_GMB-PRA_v01.pdf.

Greenberg, Miriam. *Branding New York: How a City in Crisis Was Sold to the World.* New York: Routledge, 2008.

Greene, Robert. *The 48 Laws of Power.* New York: Penguin, 1998.

Greer, Olivia J. "No Cause of Action: Video Surveillance in New York City." *Michigan Telecommunications and Technology Law Review* 18, no. 2 (2012): 589-626.

Grosz, Elizabeth. *Chaos, Territory, Art: Deleuze and the Framing of the Earth.* Durham: Duke University Press, 2008.

Grove, Kevin. "From Emergency Management to Managing Emergence: A Genealogy of Disaster Management in Jamaica." *Annals of the Association of American Geographers* 103, no. 3 (2013): 570-588.

Grove, Kevin and David Chandler. "Resilience and the Anthropocene: The Stakes of 'Renaturalising' Politics." *Resilience: International Policies, Practices and Discourses* 5, no. 2 (2016): 79-91. Doi:10.1 080/21693293.2016.1241476

Grove, Kevin. *Resilience.* Abingdon: Routledge, 2018.

Grusin, Richard, ed. *Anthropocene Feminism.* Minneapolis: University of Minnesota Press, 2017.

Gunderson, Lance. "Living with Uncertainty and Surprise." Stockholm Resilience Centre TV. Video file. January 23, 2009. https://www.youtube.com/watch?v=kqkfHjX9IsY

Gunderson, Lance H. and C.S. Holling. *Panarchy: Understanding transformations in systems of humans and nature.* Washington, DC: Island Press, 2002.

Gunderson, Lance, C.S. Holling, and Stephen Light. *Barriers and Bridges to the Renewal of Ecosystems and Institutions.* New York: Columbia University Press, 1995.

Haff, Peter K. "Technology as a Geological Phenomenon: Implications for Human Well Being." *Geological Society, London, Special Publications* 395 (2014): 301–309. https://doi.org/10.1144/SP395.4.

Hamilton, Clive. "Human Destiny in the Anthropocene." In *The Anthropocene and the Global Environment Crisis—Rethinking Modernity in a New Epoch*, edited by Clive Hamilton, Christophe Bonneuil and François Gemenne, 32-43. Abingdon: Routledge, 2013.

Hamilton, Clive. "The Delusion of the "Good Anthropocene": Reply to Andrew Revkin." June 17, 2014, https://clivehamilton.com/the-delusion-of-the-good-anthropocene-reply-to-andrew-revkin/.

Haraway, Donna. *Staying with the Trouble: Making Kin in the Chthulucene*. Durham, NC: Duke University Press, 2016.

Haraway, Donna. "Tentacular thinking: Anthropocene, Capitalocene, Chthulucene." *e-flux* 75, September 2016. https://www.e-flux.com/journal/75/67125/tentacular-thinking-anthropocene-capitalocene-chthulucene/.

Harris, Alex. "Miami Beach wants higher roads and pumps to fight sea rise. Some residents say no way." *Miami Herald*, May 16, 2018. https://www.miamiherald.com/news/local/community/miami-dade/miami-beach/article211237324.html.

Hartley, Daniel. "Against the Anthropocene." *Salvage*, 2015. http://salvage.zone/in-print/against-the-anthropocene/.

Heidegger, Martin. *Being and Time*. New York, NY: HarperCollins, 1962.

Heidegger, Martin. "Nietzsche's Word: 'God is Dead'," in *Off the Beaten Track*. Cambridge, MA: Cambridge University Press, 2002.

Henderson, Scott. "CrossFit's Explosive Affiliate Growth by the Numbers." October 23, 2018. https://morningchalkup.com/2018/10/23/crossfits-explosive-affilaite-growth-by-the-numbers/.

Herz, Jessie Cameron. "NerdFit: Why Technies Love CrossFit." *BoingBoing*, August 12, 2014. https://boingboing.net/2014/08/12/nerdfit-why-techies-love-cro.html.

Herz, Jessie Cameron. *Learning to Breathe Fire: The Rise of CrossFit and the Primal Future of Fitness*. New York: Three Rivers Press, 2014.

Higman, Barry W. *Slave Population and Economy in Jamaica, 1807-1834*. Barbados: The Press University of the West Indies, 1995.

Higman, Barry W. *A Concise History of the Caribbean*. Cambridge: Cambridge University Press, 2010.

Hoffmann, Matthew J. *Climate Governance at the Crossroads: Experimenting with a Global Response after Kyoto*. Oxford: Oxford University Press, 2011.

Holling, C.S. "Resilience and Stability of Ecological Systems." *Annual Review of Ecology and Systematics* 4, no. 1 (1973): 1-23. Doi: 10.1146/annurev.es.04.110173.000245.

Holling, C.S. "Understanding the Complexity of Economic, Ecological, And Social Systems." *Ecosystems* 4, no. 5 (2001), 390-405. http://dx.doi.org/10.1007/s10021-001-0101-5.

Holling, C.S. "From Complex Regions to Complex Worlds." *Ecology and Society* 9, no. 1 (2004): 11. http://www.ecologyandsociety.org/vol9/iss1/art11.

Holling, C.S. "Resilience Dynamics." Stockholm Resilience Centre TV. Video file. November 5, 2008. https://www.youtube.com/watch?v=FhfmaXZPKEY&t=7s

Holling, C.S. "Resilience and Life in the Arctic." *Resilience Science*, April 5, 2011. http://rs.resalliance.org/2011/04/05/resilience-and-life-in-the-arctic/

Homer-Dixon, Thomas. *The Upside of Down: Catastrophe, Creativity, and the Renewal of Civilization*. Washington, D.C.: Island Press, 2006), 228.

"How to Survive a Disaster in a Big City - According to an Urban Prepper." NBC. Video File. Accessed March 19, 2019. https://www.nbcnews.com/feature/101/video/how-to-survive-a-disaster-in-a-big-city-according-to-an-urban-prepper-1019859011805

Hui, Yuk. *The Question Concerning Technology in China: An Essay in Cosmotechnics*. Falmouth, UK: Urbanomic Media Ltd, 2016.

Invisible Committee. *To Our Friends*. Los Angeles, CA: Semiotext(e), 2014.

Invisible Committee. *Now*. South Pasadena, CA: Semiotext(e), 2017.

Jackson, Kenneth T. *Crabgrass Frontier: The Suburbanization of the United States*. New York: Oxford University Press, 1985.

Jackson, Steven J. "Rethinking Repair." In *Media Technologies: Essays on Communication, Materiality and Society*, edited by Gillespie, Tarleton, Pablo J. Boczkowski, and Kristen A. Foot, 221-239. Cambridge: MIT Press, 2014.

Jakubowski, Marcin. "Hacking the Farm with Low-Cost, Open Source Tool Designs." Interview with Don Watkins. *Opensource*, February 15, 2016. https://opensource.com/life/16/2/interview-marcin-jakubowski-open-source-ecology.

James, Eric P. and Rebecca Gill. "Neoliberalism and the Communicative Labor of CrossFit." *Communication & Sport* 6, no. 6 (2018): 703–27. Doi:10.1177/2167479517737036,

Jargon, Julie. "Too Much Coffee? Starbucks Shops Outnumber McDonald's." *Wall Street Journal*, June 7, 2018, https://www.wsj.com/articles/too-much-coffee-starbucks-shops-outnumber-mcdonalds-1528372800.

Jemisin, N.K. *The Fifth Season (The Broken Earth)*. London: Orbit Books, 2015.

Johansen, Bruce Elliott and Barbara Alice Mann, eds. *Encyclopedia of the Haudenosaunee (Iroquois Confederacy)*. Westport, CT: Greenwood Publishing Group, 2000.

Johnson, Hume, N. "Incivility: The Politics of 'People on the Margins' in Jamaica." *Political Studies* 53 (2005): 579-597. Doi: 10.1111/j.1467-9248.2005.00545.x.

Joyce, Patrick. *The Rule of Freedom: The City and Modern Liberalism*. London: Verso, 2003.

Junger, Sebastian. *Tribe: On Homecoming and Belonging*. New York: Hachette Book Group, 2016.

Kaika, Maria and Erik Swyngedouw. "Fetishizing the Modern City: The Phantasmagoria of Urban Technological Networks." *International Journal of Urban and Regional Research*, 24, no. 1 (2000): 120–138. https://doi.org/10.1111/1468-2427.00239.

Kamph, Stefan. "Fallout Fanatics." *Miami New Times*, April 26, 2012. http://digitalissue.miaminewtimes.com/article/Fallout+Fanatics/1043822/108976/article.html

Kane, Michael. "Why New York Should Become the City of Oysters Again." *New York Post*, June 21, 2014. https://nypost.com/2014/06/21/why-new-york-should-become-the-city-of-oysters-again/.

Kavanagh, Jennifer and Michael D. Rich. "An Initial Exploration of the Diminishing Role of Facts and Analysis in American Public Life." Santa Monica, CA: RAND Corporation, 2018. https://www.rand.org/pubs/research_reports/RR2314.html.

Kelly, Casey Ryan. "The Man-pocalpyse: Doomsday Preppers and the Rituals of Apocalyptic Manhood." *Text and Performance Quarterly* 36, no. 2-3, (2016): 95-114. Doi: 10.1080/10462937.2016.1158415.

Kilcullen, David. *Out of the Mountains: The Coming Age of the Urban Guerrilla.* London: C. Hurst & Co., 2016.

Klein, Naomi. *The Shock Doctrine: The Rise of Disaster Capitalism.* Toronto: A.A. Knopf Canada, 2007.

Kniveton, Dominic. "Sea Level Rise Impacts: Questioning inevitable migration." *Nature Climate* Change 7, no. 8 (2017): 548-549.

Kohso, Sabu, in Silvia Federici, George Caffentzis, Anne Waldman, Daniel de Roulet, and Sabu Kohso. *Fukushima Mon Amour.* New York: Autonomedia, 2011.

Koolhaas, Rem. *Delirious New York: A Retroactive Manifesto for Manhattan.* New York: Monacelli Press, 1997.

Korten, Tristram. "In Florida, Officials Ban Term 'Climate Change'." *Miami Herald,* March 8, 2015. https://www.miamiherald.com/news/state/florida/article12983720.html.

Kubler, Kyle. "Auto-Body." *Ultra,* August 31, 2016. http://www.ultra-com.org/project/auto-body/

Lamar, Kendrick. "HUMBLE." *Damn.* Top Dawg Entertainment, Aftermath Entertainment and Interscope Records, 2018.

Lamb, Matthew D. and Cory Hillman. "Whiners Go Home: Tough Mudder, Conspicuous Consumption, and the Rhetorical Proof of 'Fitness'." *Communication & Sport* 3, no. 1 (2015): 81–99. Doi:10.1177/2167479514521598.

Larkin, Brian. "The Politics and Poetics of Infrastructure." *Annual Review of Anthropology* 42 (2013): 327–343. https://doi.org/10.1146/annurev-anthro-092412-155522.

Latour, Bruno. "Facing Gaia: Six Lectures on the Political Theology of Nature." The Gifford Lectures on Natural Religion, Edinburgh, February 18-28, 2013a. http://www.bruno-latour.fr/sites/default/files/downloads/GIFFORD-SIX-LECTURES_1.pdf.

Latour, Bruno. "Telling Friends from Foes in the Time of the Anthropocene." In *The Anthropocene and the Global Environment Crisis—Rethinking Modernity in a New Epoch*, edited by Clive Hamilton, Christophe Bonneuil and François Gemenne, 145-155. London, Routledge, 2013b.

Latour, Bruno. *Facing Gaia: Eight Lectures on the Mew Climatic Regime*. Cambridge, UK; Medford, MA: Polity Press, 2017.

Latour, Bruno, Isabelle Stengers, Anna Tsing, and Nils Bubandt. "Anthropologists are Talking—About Capitalism, Ecology, and Apocalypse." *Ethnos* 83, no. 3 (2018): 587-606. Doi: 10.1080/00141844.2018.1457703.

Law, John. "What's Wrong with a One World World?" *Distinktion: Journal of Social Theory* 16, no. 1 (2015): 126-139. Doi: 10.1080/1600910X.2015.1020066.

Levinas, Emmanuel. *Totality and Infinity: An Essay in Exteriority*. Pittsburgh, Pennsylvania: Duquesne University Press, 1969.

Levine, Philip. "Philip Levine for Mayor of Miami Beach 'Paddle'." Campaign ad, video, 2013. https://www.youtube.com/watch?v=N9niAnh9KZw.

Lewis, Simon L. and Mark A. Maslin. "Defining the Anthropocene." *Nature* 519 no. 7542 (2015): 171-180. Doi: 10.1038/nature14258.

Limberg, Peter and Conor Barnes. "The Memetic Tribes of Culture War 2.0." *Medium*, September 13, 2018. https://medium.com/s/worldwide-wtf/memetic-tribes-and-culture-war-2-0-14705c43f6bb.

Lorimer, Jamie. *Wildlife in the Anthropocene: Conservation after Nature*. Minneapolis: University of Minnesota Press, 2015.

Lorimer, Jamie and Clemens Driessen. "Wild Experiments at the Oostvaardersplassen: Rethinking Environmentalism in the Anthropocene." *Transactions of the Institute of British Geographers* 39, no. 2 (2013): 169–181. Doi:10.1111/tran.12030.

Lorimer, Jamie, Chris Sandom, Paul Jepson, Chris Doughty, Maan Barua, and Keith J. Kirby. "Rewilding: Science, Practice, and Politics." *Annual Review of Environment and Resources* 40 (2015): 39–6.

Lourie Harrison, Ariane. *Architectural Theories of the Environment: Posthuman Territory*. New York: Routledge, 2013.

Lovins, Amory B. and L. Hunter Lovins. *Brittle Power: Energy strategy for National Security.* Denver: Brick House Pub Company, 1982.

Malm, Andreas and Alf Hornborg. "A Geology of Mankind? A Critique of the Anthropocene Narrative." *The Anthropocene Review* 1, no. 1 (2014): 62–59.

Marcus, Aubrey. *Own the Day, Own Your Life.* London: Harper Collins, 2018.

Marks, Robert. *The Origins of the Modern World: A Global and Ecological Narrative from the Fifteenth to the Twenty-First Century.* Lanham, Md: Rowman & Littlefield, 2007.

Marres, Noortje. "Why We Can't Have Our Facts Back." *Engaging Science, Technology, and Society* 4 (2018): 423-443. Doi:10.17351/ests2018.188.

Marris, Emma. "How a Few Species are Hacking Climate Change." *National Geographic,* May 6, 2014. https://news.nationalgeographic.com/news/2014/05/140506-climate-change-adaptation-evolution-coral-science-butterflies/.

Marx, Karl and Friedrich Engels. *The German Ideology.* Moscow: Progress Publishers, 1964.

Massumi, Brian. "National Enterprise Emergency: Steps Toward an Ecology of Powers." *Theory, Culture & Society* 26, no. 6 (2009): 153-185.

McCastle, Michael. "Imagination: The Mind's Contribution to Peak Performance." *Breaking Muscle.* Accessed March 19, 2019. https://breakingmuscle.com/fitness/imagination-the-minds-contribution-to-peak-performance

McClatchy. "Everglades Python Challenge Wraps Up." *Miami Herald,* February 15, 2016. https://www.miamiherald.com/news/local/environment/article60450246.html

McDougall, Christopher. *Natural Born Heroes: Mastering the Lost Secrets of Strength and Endurance.* New York: Vintage, 2016.

Meaney, Thomas. "Populist Insurgency." *New Yorker,* February 26, 2018. https://www.newyorker.com/magazine/2018/02/26/a-celebrity-philosopher-explains-the-populist-insurgency

Meeks, Brian. *Envisioning Caribbean Futures: Jamaican Perspectives.* Kingston: University of West Indies Press, 2007.

Meerow, Sara A., Joshua P. Newell, and Melissa Stults. "Defining Urban Resilience: A Review." *Landscape and Urban Planning* 147 (2016): 38–49.

Meschino, Patricia. "Is Chronixx Reggae's Next Big Thing? Chris Blackwell Thinks So!" *The Foundation Radio Media Network*, September 25, 2013, http://www.clintonlindsay.com/2013/09/25/is-chronixx-reggaes-next-big-thing-chris-blackwell-thinks-so/

Mitchell, Richard G. *Dancing at Armageddon: Survivalism and Chaos in Modern Times*. Chicago: University of Chicago Press, 2002.

Mitchell, Timothy. *Colonizing Egypt*. Berkeley, CA: University of California Press, 1988.

Mitchell, Timothy. *Carbon Democracy: Political Power in the Age of Oil*. London and New York: Verso, 2013.

Mitropoulos, Angela. *Contract & Contagion: From Biopolitics to Oikonomia*. New York: Minor Compositions, 2012.

Moore, Jason, ed., *Anthropocene or Capitalocene? Nature, History, and the Crisis of Capitalism*. Oakland, CA: PM Press, 2015.

Morton, Tim. *Hyperobjects: Philosophy and Ecology After the End of the World*. Minneapolis: University of Minnesota Press, 2013.

Mrázek, Rudolf. *Engineers of Happy Land: Technology and Nationalism in a Colony*. Princeton, NJ: Princeton University Press, 2002.

Nagle, Angela. *Kill All Normies: Online Culture Wars from 4chan and Tumbler to Trump and the Alt-Right*. Zero Books, 2017.

NATO Infrastructure Committee. *50 Years of Infrastructure: NATO Security Investment Programme is the Sharing of Roles, Risks, Responsibilities, Costs and Benefits*. Brussels, Belgium: NATO Infrastructure Committee, 2001.

Neel, Phil. *Hinterland: America's New Landscape of Class and Conflict*. London: Reaktion Books, 2018.

Nehamas, Nicholas. "Miami's downtown building boom drawing to a close." *Miami Herald*, October 14, 2015. http://www.miamiherald.com/news/business/real-estate-news/article39189630.html.

Neimark, Benjamin, John Childs, Andrea J. Nightingale, Connor Joseph Cavanagh, Sian Sullivan, Tor A. Benjaminsen, Simon Batterbury, Stasja Koot, and Wendy Harcourt. "Speaking Power to 'Post-Truth': Critical Political Ecology and the New Authoritarianism." *Annals of the American Association of Geographers* 109, no. 2 (2019): 613-623.

Nelson, Sara Holiday. "Resilience and the neoliberal counter-revolution: from ecologies of control to production of the common." *Resilience: International Policies, Practices and Discourses* 2, no. 1 (2014): 1-17.

Nerem, Robert Steven, B. D. Beckley, John T. Fasullo, Benjamin Hamlington, D. Masters, G. T. Mitchum. "Climate-Change–Driven Accelerated Sea-Level Rise." *Proceedings of the National Academy of Sciences* 115, no. 9 (2018): 2022-2025. Doi: 10.1073/pnas.1717312115.

New York City Office of Emergency Management. "Meet Ready Girl." 2019. https://www1.nyc.gov/site/em/ready/ready-girl.page.

New York City Panel on Climate Change. "Climate Risk Information 2013: Observations, Climate Change Projections, and Maps," Report, 2013. http://www.nyc.gov/html/planyc2030/downloads/pdf/npcc_climate_risk_information_2013_report.pdf.

Niaah, Sonjah Stanley. *Dancehall: From Slave Ship to Ghetto*. Ottawa: University of Ottawa Press, 2010.

Nietzsche, Friedrich. *The Will to Power*. New York: Random House, 1967.

Nietzsche, Friedrich. *Twilight of the Idols, or, How to Philosophize with a Hammer*. New York: Oxford University Press, 1998.

Noble, David. *Forces of Production: A Social History of Industrial Automation*. Oxford: Oxford University Press, 1984.

Norbert, Sébastian, Julien Rebotier, Cloé Vallette, Christine Bouisset and Sylvie Clarimont. "Resilience for the Anthropocene? Shedding Light on The Forgotten Temporalities Shaping Post-Crisis Management in the French Sud Ouest." *Resilience: International Policies, Practices and Discourses* 5, no. 3 (2017): 145–160.

Olsson, Per, Victor Galaz, and Wiebren J. Boonstra. "Sustainability Transformations: A Resilience Perspective." *Ecology and Society* 19, no. 4: 1. http://dx.doi.org/10.5751/ES-06799-190401.

Olsson, Per, Carl Folke, Thomas Hahn. "Social-Ecological Transformation for Ecosystem Management: The Development of Adaptive Co-Management of a Wetland Landscape in Southern Sweden." *Ecology and Society* 9, no. 2 (2004). Doi: 10.5751/ES-00683-090402.

Open Source Ecology. "Machines: Global Village Construction Set." *Open Source Ecology.* Accessed March 19, 2019. https://www.opensourceecology.org/gvcs/

Osborne, Thomas. "Security and Vitality: Drains, Liberalism, and Power in the Nineteenth Century." In *Foucault and Political Reason: Liberalism, Neo-liberalism, and Rationalities of Government*, edited by Andrew Barry, Nikolas Rose, & Thomas Osborne, 99-121. Chicago: University of Chicago Press, 1996.

Otter, Chris. "Making Liberalism Durable: Vision and Civility in the Late Victorian City." *Social History* 27, no. 1 (2002): 1-15.

Pakalolo. "The Phenomenon that Cannot be Spoken in Florida Continues as Salt Water Intrusion Moves Inland." *Daily Kos,* March 19, 2015. http://www.dailykos.com/story/2015/03/19/1372031/-The-phenomenon-that-can-not-be-spoken-in-Florida-continues-as-salt-water-intrusion-moves-inland

Papadopoulos, Dimitris. *Experimental Practice: Technoscience, Alterontologies, and More-Than-Social Movements.* Durham, NC: Duke University Press, 2018.

Patterson, James T. *Grand Expectations: The United States, 1945–1974.* Oxford and New York: Oxford University Press, 1996.

Pawlowski, Thaddeus and Michelle Mueller. "The Resilience Accelerator—Getting up to Speed in Southeast Florida." *100 Resilient Cities* blog. November 20, 2018. https://www.100resilientcities.org/resilience-accelerator-getting-up-to-speed-southeast-florida/.

Perkol-Finkel, Shimrit and Ido Sella. "Ecologically Active Concrete for Coastal and Marine Infrastructure: Innovative Matrices and Designs." *Proceeding of the 10th ICE Conference: From Sea to Shore - Meeting the Challenges of the Sea,* 2014. Doi: 10.1680/fsts597571139.

Portal, Ido. "Ido Portal: Just Move." Brian Rose and London Real. Video File. 2014. https://londonreal.tv/e/ido-portal-just-move/

Portes, Alejandro and Ariel C. Armony. *The Global Edge: Miami in the 21st Century.* Oakland, CA: University of California Press, 2018.

Povinelli, Elizabeth A. "The Persistence of Hope: Critical Theory and Enduring in Late Liberalism." In *Theory After 'Theory,'* edited by Jane Elliott and Derek Attridge. Abingdon: Routledge, 2011.

Povinelli, Elizabeth A. *Geontologies: A Requiem to Late Liberalism.* Durham, N.C.: Duke University Press, 2016.

Powell, Keleshia. "Chronixx Prepared to be Judged for 'Waste Man' Post on Instagram." *Jamaica Observer,* April 10, 2015. http://www.jamaicaobserver.com/news/Chronixx-removes--waste-man--post-from-Instagram-after-harsh-criticisms.

Proteje featuring Chronixx. "No Guarantee." From Proteje, *A Matter of Time.* In.Digg.Nation Collective/Overstand Entertainment under license to Easy Star Records, 2018.

"Raise Up: The World is Our Gym." Video. Directed by B. Rain Bennett. Flying Flounder Productions, Red Bull Media House. 2017. https://www.redbull.com/int-en/tv/film/AP-1RDEMV9D12111/raise-up.

Rebuild by Design. "The Rebuilders." Video file. 2014. https://vimeo.com/90825595.

Rebuild by Design. "Promoting resilience post-Sandy through innovative planning and design." Rebuild by Design: Hurricane Sandy Regional Planning and Design Competition. Design brief. June 21, 2013. http://portal.hud.gov/hudportal/documents/huddoc?id=REBUILDBYDESIGNBrief.pdf.

Reel, Monte. "How to Rebuild Puerto Rico." *Bloomberg Businessweek,* December 27, 2017. https://www.bloomberg.com/news/features/2017-12-14/how-to-rebuild-puerto-rico.

Regalado, Thomás. "We cannot allow this to become the new 'normal'." Twitter. August 2, 2017.

Reiter, Bernd. *Constructing the Pluriverse: The Geopolitics of Knowledge.* Durham, NC: Duke University Press, 2018.

Revkin, Andrew. "Varied Views (Dark, Light, in Between) of Earth's Anthropocene Age." *Dot Earth New York Times* blog, July 15, 2015. https://dotearth.blogs.nytimes.com/2015/07/15/varied-views-dark-light-in-between-of-earths-anthropocene-age.

Robb, John. *Brave New War: The Next Stage of Terrorism and the End of Globalization.* Hoboken, NJ: Wiley, 2008.

Rockefeller Foundation. "100 Resilient Cities Announces Global Summit—Largest Ever Gathering of Urban Resilience Experts." Press release, June 23, 2017. https://www.100resilientcities.org/100-resilient-cities-announces-largest-ever-gathering-of-urban-resilience-e/.

Rockefeller Foundation. "The Resilience Age." 2016. Video. https://www.youtube.com/watch?v=w-wDyhewNZ0.

Rockström, Johan. "Let the Environment Guide our Development." Video. TED Talk. TEDGlobal. 2010. https://www.ted.com/talks/johan_rockstrom_let_the_environment_guide_our_development.

Rockström, Johan. "Beyond the Anthropocene." *Stockholm Resilience Centre*, 2017. Video. http://www.stockholmresilience.org/research/research-news/2017-02-16-wef-2017-beyond-the-anthropocene.html.

Rockström, Johan, Will Steffen, Kevin Noone, Åsa Persson, F. Stuart III Chapin, Eric Lambin, Timothy M. Lenton, Marten Scheffer, Carl Folke, Hans Joachim Schellnhuber, Björn Nykvist, Cynthia A. de Wit, Terry Hughes, Sander van der Leeuw, Henning Rodhe, Sverker Sörlin, Peter K. Snyder, Robert Costanza, Uno Svedin, Malin Falkenmark, Louise Karlberg, Robert W. Corell, Victoria J. Fabry, James Hansen, Brian Walker, Diana Liverman, Katherine Richardson, Paul Crutzen, and Jonathan Foley. "Planetary Boundaries: Exploring the Safe operating Space for Humanity." *Ecology and Society* 14, no. 2 (2009): 32. http://www.ecologyandsociety.org/vol14/iss2/art32/.

Rodin, Judith. "Realizing the Resilience Dividend." *Rockefeller Foundation*, January 22, 2014. https://www.rockefellerfoundation.org/blog/realizing-resilience-dividend/.

Ronfelt, David. "IN SEARCH OF HOW SOCIETIES WORK: Tribes: The First and Forever Form." RAND Corporation Working Paper WR-433-RPC, 2007. https://www.rand.org/pubs/working_papers/WR433.html.

Rosenzweig, Cynthia and William Solecki. "Hurricane Sandy and Adaptation Pathways in New York: Lessons from a First-Responder City." *Global Environmental Change* 28, no. 1 (2014): 395-408. Doi:10.1016/j.gloenvcha.2014.05.003.

Rosenzweig, Cynthia and William Solecki. "Building Climate Resilience in Cities: Lessons from New York." *The Conversation*, January 22, 2016. https://theconversation.com/building-climate-resilience-in-cities-lessons-from-new-york-52363.

Roskind, Robert. *Rasta Heart: A Journey into One Love.* Blowing Rock, N.C.: One Love Press, 2001.

Rosol, Christoph, Sara Nelson, and Jürgen Renn. "In the Machine Room of the Anthropocene." *The Anthropocene Review* 4, no. 1 (2017): 2–8. https://doi.org/10.1177/2053019617701165.

Ruggeri, Amanda. "Miami's Fight Against Rising Seas." *BBC Future Now,* April 4. 2017. http://www.bbc.com/future/story/20170403-miamis-fight-against-sea-level-rise.

Sahlins, Marshall. *The Western Illusion of Human Nature.* Chicago, Prickly Paradigm Press, 2008.

Savransky, Martin. *The Adventure of Relevance: An Ethics of Social Inquiry.* New York, NY: Palgrave Macmillan, 2016.

SCAPE / Landscape Architecture. *Rebuild by Design/ Living Breakwaters. IP Report, Staten Island and Raritan Bay.* New York: Rebuild by Design, 2013.

Schmidt, Eric and Jared Cohen. *The New Digital Age: Transforming Nations, Businesses, and Our Lives.* New York: Vintage Books, 2014.

Schmitt, Carl. *The Nomos of the Earth in the International Law of the Jus Publicum Europaeum.* New York: Telos, 2003.

Schneider, Jane and Ida Susser. *Wounded Cities: Destruction and Reconstruction in a Globalized World.* Oxford; New York: Berg, 2003.

Schor, Juliet. *The Overworked American: The Unexpected Decline of Leisure.* New York, NY: Basic Books, 1991.

Schürmann, Reiner. ""What Must I Do?' at the End of Metaphysics: Ethical Norms and the Hypothesis of a Historical Closure." In *Phenomenology in a Pluralistic Context,* edited by William I. McBride and Calvin O. Schrag, 49-64. Binghamton: SUNY Press, 1984.

Schürmann, Reiner. *Heidegger on Being and Acting: From Principles to Anarchy.* Bloomington: Indiana University Press, 1987.

Scott, David. *Refashioning Futures: Criticism after Postcoloniality.* Princeton, N.J: Princeton University Press, 1999.

Scott, David and Stuart Hall. "David Scott by Stuart Hall." *BOMB Magazine,* January 1, 2005. https://bombmagazine.org/articles/david-scott/.

Scott, David. *Conscripts of Modernity: The Tragedy of Colonial Enlightenment*. Durham, N.C.: Duke University Press, 2004.

Scott, David. "Political Rationalities of the Jamaican Modern." *Small Axe* 7, no. 2 (2003): 1-22.

Scranton, Roy. *Learning to Die in the Anthropocene: Reflections on the End of a Civilization*. San Francisco, CA: City Lights Publishers, 2015.

SeArc - Ecological Marine Consulting. "About SeArc." 2010. http://www.searc-consulting.com/home.yecms/index.

Serwer, Jesse. "GEN F: Chronixx. Chronixx is leading Jamaican music's roots revival." *THE FADER*. September 20, 2013. https://www.thefader.com/2013/09/20/gen-f-chronixx.

Shishido, Takako, Ayumi Hirai, Sabu Kohso, Yuko Tonohira, "Voluntary Evacuation: A New Form of Struggle: A Conversation with Takako Shishido by Todos Somos Japon." June 23, 2012. https://jfissures.wordpress.com/2013/01/14/voluntary-evacuation-a-new-form-of-struggle-a-conversation-with-takako-shishido-1/.

Simone, AbdouMaliq. *For the City Yet to Come: Changing African Life in Four Cities*. Durham: Duke University Press, 2004.

Sloterdijk, Peter. *Bubbles: Spheres Volume 1: Microsphereology*. Los Angeles, CA: Semiotext(e), 2011.

Sloterdijk, Peter. *You Must Change Your Life*. Cambridge, UK: Polity Press, 2013.

Smith, Neil. "The Production of Nature." In *Futurenatural: Nature, Science, Culture*, edited by Jon Bird, Barry Curtis, Melinda Mash, Tim Putnam, George Robertson, and Lisa Tickner, 35–54. London: Routledge, 1996.

Smith, Neil. *Uneven Development: Nature, Capital, and the Production of Space*. Athens, GA and London: University of Georgia Press, 1990.

Smith, Neil. *The New Urban Frontier: Gentrification and the Revanchist City*. London and New York: Routledge, 1996.

Southeast Florida Regional Climate Change Compact (SFRCCC). "Unified sea level rise Projection for southeast Florida." October 2015. http://www.southeastfloridaclimatecompact.org/wp-content/uploads/2015/10/2015-Compact-Unified-Sea-LevelRise-Projection.pdf.

Spanger-Siegfried, Erika. Melanie Fitzpatrick, and Kristina Dahl. "Encroaching Tides: How Sea Level Rise and Tidal Flooding Threaten US East And Gulf Coast Communities Over the Next 30 Years." Cambridge, MA: Union of Concerned Scientists, 2014. http://www.ucsusa.org/encroachingtides.

Steffen, Will, Wendy Broadgate, Lisa Deutsch, Owen Gaffney, and Cornelia Ludwig. "The Trajectory of the Anthropocene: The Great Acceleration." *The Anthropocene Review* 2, no. 1 (2015), 81–98. Doi:10.1177/2053019614564785.

Steffen, Will, Paul J. Crutzen, and John R. McNeill. "The Anthropocene: Are Humans Now Overwhelming the Great Forces of Nature?" *AMBIO: A Journal of the Human Environment* 36, no. 8 (2007): 614-621. Doi: 10.1579/0044-7447(2007)36[614:TAAHNO]2.0.CO;2.

Steffen, Will, Katherine Richardson, Johan Rockström, Sarah E. Cornell, Ingo Fetzer, Elena M. Bennett, Reinette Biggs, Stephen R. Carpenter, Wim De Vries, Cynthia A. De Wit, Carl Folke, Dieter Gerten, Jens Heinke, Georgina M. Mace, Linn M. Persson, Veerabhadran Ramanathan, Belinda Reyers, Sverker Sörlin. "Planetary Boundaries: Guiding Human Development on a Changing Planet." *Science* 347, no. 6223 (2015): 1259855-1-1259855-10. Doi:10.1126/science.1259855.

Steffen, Will, Johan Rockström, Katherine Richardson, Timothy M. Lenton, Carl Folke, Diana Liverman, Colin P. Summerhayes, Anthony D. Barnosky, Sarah E. Cornell, Michel Crucifix, Jonathan F. Donges, Ingo Fetzer, Steven J. Lade, Marten Scheffer, Ricarda Winkelmann, and Hans Joachim Schellnhuber. "Trajectories of the Earth System in the Anthropocene." *Proceedings of the National Academy of Sciences* 115, no. 33 (2018): 8252-8259. https://doi.org/10.1073/pnas.1810141115.

Stein, Kate. "'We're a Living Laboratory:' Miami Beach Works on Resiliency as Businesses Face Flooding." *WLRN*, October 17, 2016. http://www.wlrn.org/post/were-living-laboratory-miami-beach-works-resiliency-businesses-face-flooding.

Stengers, Isabelle. *In Catastrophic Times: Resisting the Coming Barbarism*. London: Open Humanities Press, 2015.

Strauss, Neil. "Brock Pierce: The Hippie King of Cryptocurrency." *Rolling Stone*, July 26, 2018. https://www.rollingstone.com/culture/culture-features/brock-pierce-hippie-king-of-cryptocurrency-700213/.

Sullivan, Andrew. "The Abyss of Hate vs. Hate." *New York Magazine*, January 25, 2019. http://nymag.com/intelligencer/2019/01/andrew-sullivan-the-abyss-of-hate-versus-hate.html

Taleb, Nassim Nicholas. *Skin in the Game: Hidden Asymmetries in Daily Life*. New York: Random House, 2018.

Tarr, Béla. Interview with Vladan Petkovic. *Cineuropa*, April 3, 2011. https://cineuropa.org/en/interview/198131/.

"They Call Us Mohawks: Ganienkeh: A Way of Life." Turtle Island Trust Documentary, 2013. https://www.youtube.com/watch?v=9NqeyFIWnlc.

Tollefson, Jeff. "Natural Hazards: New York vs. the Sea." *Nature*, February 13, 2013. http://www.nature.com/news/natural-hazards-new-york-vs-the-sea-1.12419.

Trump, Donald (@realDonaldTrump). 2017. "Writing my inaugural address at the Winter White House, Mar-a-Lago, three weeks ago. Looking forward to Friday. #Inauguration." Twitter, January 18, 2017, 12:33 PM. https://twitter.com/realDonaldTrump/status/821772494864580614.

Tsing, Anna. *The Mushroom at the End of the World: On the Possibility of Life in Capitalist Ruins*. Princeton, NJ: Princeton University Press, 2016.

Turnheim, Bruno, Paula Kivimaa, and Frans Berkhout (Eds.), *Innovating Climate Governance: Moving Beyond Experiments*. Cambridge: Cambridge University Press, 2018.

Union of Concerned Scientists. "Encroaching Tides in Miami-Dade County, Florida. Fact Sheet: Sea Level Rise and Tidal Flooding Along the East Coast." 2016. https://www.ucsusa.org/sites/default/files/attach/2016/04/miami-dade-sea-level-rise-tidal-flooding-fact-sheet.pdf.

Urban Land Institute (ULI). "Stormwater Management and Climate Adaptation Review. An Urban Land Institute Advisory Services Panel Report." Miami Beach, FL, April 16-19, 2018. http://www.mbrisingabove.com/wp-content/uploads/2018/04/Miami-Beach_Panel-Report_lo-res.pdf.

Urken, Ross Kenneth. "What Rasta Taught Chronixx." *FADER*, March 6, 2017. http://www.thefader.com/2017/03/06/what-rasta-taught-chronixx.

VanderMeer, Jeff. *Borne*. Toronto, Ontario, Canada: HarperCollins Publishers Ltd, 2017.

Wacquant, Loïc. "Class, Race and Hyperincarceration in Revanchist America." *Daedalus* 139, no. 3 (2010): 74-90.

Wakefield, Stephanie. "Man in the Anthropocene (as portrayed by the film Gravity)." *May* 13 (2014). http://www.mayrevue.com/en/lhomme-de-lanthropocene-tel-que-depeint-dans-le-film-gravity/.

Wakefield, Stephanie. "Inhabiting the Anthropocene Back Loop." *Resilience: International Policies, Practices and Discourses* 6, no. 1 (2017): 1-18. Doi: 10.1080/21693293.2017.1411445

Wakefield, Stephanie. "Infrastructures of Liberal Life: From Modernity and Progress to Resilience and Ruins." *Geography Compass* 12, no. 7 (2018). Doi: 10.1111/gec3.12377.

Wakefield, Stephanie and Bruce Braun. "Governing the Resilient City." *Environment and Planning D: Society and Space* 32, no. 1 (2014): 4-11.

Wakefield, Stephanie and Bruce Braun. "Inhabiting the Post-Apocalyptic City." *Society and Space Open*, 2014. https://societyandspace.org/2014/02/11/inhabiting-the-postapocalytic-city-bruce-braun-and-stephanie-wakefield/.

Wakefield, Stephanie and Bruce Braun. "Oystertecture: infrastructure, profanation and the sacred figure of the human." In *Infrastructure, Environment, and Life in the Anthropocene*, edited by Kregg Hetherington, 193-215. Durham, NC: Duke University Press, 2019.

Walker, Brian, Lance Gunderson, Ann Kinzig, Carl Folke, Steve Carpenter, and Lisen Schultz. "A handful of heuristics and some propositions for understanding resilience in social-ecological systems. *Ecology and Society* 11, no. 1 (2006): 13. http://www.ecologyandsociety.org/vol11/iss1/art13/.

Walker, Brian, C. S. Holling, Stephen R. Carpenter, and Ann Kinzig, "Adaptability and Transformability in Social-Ecological Systems." *Ecology and Society* 9, no. 2 (2004): 5. http://www.ecologyandsociety.org/vol9/iss2/art5/.

Walker, Brian and David Salt. *Resilience Thinking: Sustaining Ecosystems and People in a Changing World*. Washington, DC: Island Press, 2006.

Walker, Brian and David Salt. *Resilience Practice*. Washington, DC: Island Press, 2012.

Wang, Christine. "How a Health Nut Created the World's Biggest Fitness Trend." *CNBC*, April 5, 2016. https://www.cnbc.com/2016/04/05/how-crossfit-rode-a-single-issue-to-world-fitness-domination.html.

Waters, Colin, Jan Zalasiewicz, Mark Williams, Michael Ellis, and Andrea Snelling. "A Stratigraphical Basis for the Anthropocene?" *Geological Society, London, Special Publications*, 395 (2014): 1–21. https://doi.org/10.1144/SP395.18.

Wdowinski, Shimon, Ronald Bray, Ben P. Kirtman, and Zhaohua Wu. "Increasing Flooding Hazard in Coastal Communities Due to Rising Sea Level: Case study of Miami Beach, Florida." *Ocean & Coastal Management* 126 (2016): 1-8. Doi: 10.1016/j.ocecoaman.2016.03.002

Weisman, Alan. *The World Without Us*. New York: Thomas Dunne Books/St. Martin's Press, 2012.

Yusoff, Kathryn. *A Billion Black Anthropocenes or None*. Minneapolis: University of Minnesota Press, 2019.

Zalasiewicz, Jan, Will Steffen, Reinhold Leinfelder, Mark Williams and Colin Waters. "Petrifying Earth Process: the stratigraphic imprint of key Earth System parameters in the Anthropocene." *Theory, Culture & Society* 34 (2017): 83-104.

Zalasiewicz, Jan, Colin N. Waters, Mark Williams, Anthony D. Barnosky, Alejandro Cearreta, Paul Crutzen, Erle Ellis, Michael A. Ellis, Ian J. Fairchild, Jacques Grinevald, Peter K. Haff, Irka Hajdas, Reinhold Leinfelder, John McNeill, Eric O. Odada, Clement Poirier, Daniel Richter, Will Steffen, Colin Summerhayes, James P.M. Syvitski, Davor Vidas, Michael Wagreich, Scott L. Wing, Alexander P. Wolfe, An Zhishengw, Naomi Oreskes. "When did the Anthropocene begin? A Mid-Twentieth Century Boundary Level is Stratigraphically Optimal." *Quaternary International* 383 (2015): 196–203. DOI: 10.1016/j.quaint.2014.11.045.

Zalasiewicz, Jan and Mark Williams. *The Goldilocks Planet: The 4 Billion Year Story Of Earth's Climate*. Oxford: Oxford University Press, 2012.

Zalasiewicz, Jan, Mark Williams, Colin N. Waters, Anthony D. Barnosky, Peter Haff, "The Technofossil Record of Humans," *The Anthropocene Review* 1, no. 1 (2014): 34–43

Zavestoski, Stephen. "Fast Tracking Climate Adaptation: Tapping Our Natural Tendency to Experiment." *Our Place on Earth*, 2014. https://www.ourplaceonearth.org/blog?category=Guest+Blogger.

Zellmer, Sandi and Lance Gunderson. "Why Resilience May Not Always Be a Good Thing: Lessons in Ecosystem Restoration from Glen Canyon and the Everglades." *Nebraska Law Review* 87, no. 4 (2009): 893. https://ssrn.com/abstract=1434386.

Zolli, Andrew. "Learning to Bounce Back." *New York Times*, November 2, 2012. http://www.nytimes.com/2012/11/03/opinion/forget-sustainability-its-about-resilience.html.

www.ingramcontent.com/pod-product-compliance
Lightning Source LLC
Chambersburg PA
CBHW030855170426
43193CB00009BA/616